Praise for *Science of Parenthood*

"Educational and entertaining! Norine Dworkin-McDaniel and Jessica Ziegler are like the Bill Nye and Neil deGrasse Tyson of parenting . . . minus the fancy neckties."

—**DAVID VIENNA,** author of *Calm The F*ck Down*

"Adapt or die. Those are pretty much the only two real choices a person has once he or she becomes a parent. Because the only things we know for sure about this life are that it changes, and it ends. *Science of Parenthood* supplies parents with a heap of what is necessary to adapt and adapt some more and keep on adapting: Humor. Because if you don't laugh, you'll die. No, I didn't mean to say Cry. Obviously there is a lot of crying in life. And poop. Lots of poop. But if you don't laugh your way through this emotionally and physically messy world of family life, you won't survive. So, if you don't want to die, you had better buy this book."

—**NICOLE KNEPPER,** author of *Moms Who Drink and Swear*

"Move over *What to Expect When You're Expecting*. *Science of Parenthood* has the real truth about the joys, challenges, and completely unpredictable moments of parenting. Equal parts relatable, honest, accurate, and laugh-out-loud funny, *Science of Parenthood* is a must-read for anyone who has ever given birth, is pregnant, or is even thinking of having a baby. And if this last one is you, good luck. You'll need it."

—**TRACY BECKERMAN,** nationally syndicated columnist
and author of *Lost in Suburbia*

"For these two ladies, parenting IS a science. Luckily, they put plenty of humor with it, so the class clowns like me can actually understand just what the heck they're talking about!"

—**JEN MANN,** *New York Times* bestselling author of *People I Want to Punch in the Throat*

"Hysterical, informative, and therapeutic."

—**JASON GOOD,** author of *Rock, Meet Window*

"My only complaint with this book is that it didn't exist before I had kids. If you're one of the lucky ones who can get your hands on *Science of Parenthood* before your baby is born, it should be #1 on your wish list, before bassinets and burp cloths and those freakish contraptions that let you suck the snot out of its nose. And if you already have kids, get this book for the laughs and the assurances that, yes, the crazy, ridiculous miracle you've been going through is all happening exactly as it's supposed to."

—**JERRY MAHONEY,** author of *Mommy Man*

SCIENCE OF PARENTHOOD

Thoroughly Unscientific Explanations for Utterly Baffling Parenting Situations

Norine Dworkin-McDaniel & Jessica Ziegler

SHE WRITES PRESS

Published 2015
Printed in China
ISBN: 978-1-63152-947-4
Library of Congress Control Number: 2015937963

For information, address:
She Writes Press
1563 Solano Ave #546
Berkeley, CA 94707

She Writes Press is a division of SparkPoint Studio, LLC.

Dedicated to our sons, Holden and Fletcher,

who first inspired us to ask the question that every scientist starts with:

What the hell?!?

"If we couldn't laugh at ourselves, that would be the end of everything."

- Niels Bohr
Physicist

Contents

How being a parent is like being a Scientist

You have mixed together disparate genes . . . and the outcome remains uncertain.

You're surrounded by egomaniacs.

There is always a shiny new toy that must be had at all costs.

Your assumptions are routinely disproven.

You are typically underfunded and

waaaay over budget.

You are in constant, furious competition with your peers.

Your peers are only too happy to point out the flaws in your approach—and how they'd do things more effectively.

You are frequently obliged to explain yourself to people who have absolutely no clue what you're talking about.

Eureka moments are typically followed by setbacks and utter confusion.

Your best work won't be appreciated until you are old, gray . . . and possibly dead.

Solve for *Why?*

You're smart. You're savvy. You know stuff. Like, say, the difference between Syrah and Petit Syrah, and why grading on a curve is awesome. You may even know why fools fall in love and what the electoral college is. (Just kidding, *no one* knows what that is.)

But if you're reading this book, chances are you've been utterly flummoxed by one of life's great questions: *Why the hell is my kid doing X, Y, or Z? And please, dear God, is there a way to make him stop? Like RIGHT NOW?!?*

Welcome to the club: Bewildered Parents R Us. Come in. Sit down. You'll probably be here for the next eighteen to thirty-five years, so you might as well get comfy.

When people ask hard questions like *Why are we here?* they're called philosophers.

When people ask hard questions like *What is the universe?* they're called scientists.

And when people ask hard questions like *How can someone pass up a delicious home-cooked meal in favor of gobbling down his own boogers?* they're called parents.

We've discovered that parents are a lot like scientists. Okay, so maybe we don't have nifty gizmos like the Large Hadron Collider in our garages. Or a stash of Plutonium-238 in our pantries. But just like our buddies on the front lines of science—Galileo and Newton, Einstein and Hawking, Masters and Johnson—we always have to solve for *y*. Or, rather: *why?*

Why do children grow so fast, yet Candy Land drags on . . . so . . . s-l-o-w-l-y? Why do kids only puke when you're the sole person around to mop up? Why are rectangles the "wrong" shape for PBJs? Why are kids deaf to demands to pick up their toys, but able to hear the faint tinkle of the ice cream truck five miles away? And, seriously, WHY is Caillou such a whiny little bastard?

We seek those answers too. Which is exactly why we've dug deep into the core sciences—biology, chemistry, physics, and mathematics—to find them. And like the legions of theoreticians before us who started their quests for knowledge with the simple question *Why?*, we've come up with some intriguing hypotheses as to why kids *and* parents—not you, of course, we mean *other* parents—do the perplexing, absolutely infuriating things they do.

We'll let history judge just how right we are. Meanwhile, we'll be by the phone, awaiting our call from the Nobel Committee.

"For each of our actions, there are only consequences."

- James Lovelock

Environmentalist

Biology

BIOLOGY LESSONS

Biology—from the Greek words *bios* and *logos*—is the study of life. Beautiful, wondrous, frequently brutal life. Particularly brutal for the new parent, or the seasoned parent starting over with a newborn, whose life has been completely upended and shaken hard like one of James Bond's signature martinis. (If you like that metaphor, go grab a cocktail shaker and keep it handy. You'll need it. Routinely.)

No doubt you noticed the initial changes starting in pregnancy, as appetite was sacrificed to the body's whim for any foodstuff that didn't produce all-day nausea or vomiting. Next, the body—expanding beyond all recognition. (Bigger boobs you expected, but *bigger feet?*) And then, like dominoes, went modesty, privacy, and any hope of personal grooming.

Before you knew it, you were indentured to a small, capricious dictator who's hell-bent on pushing you right over the edge.

That's life.

Should You Have Another Baby?

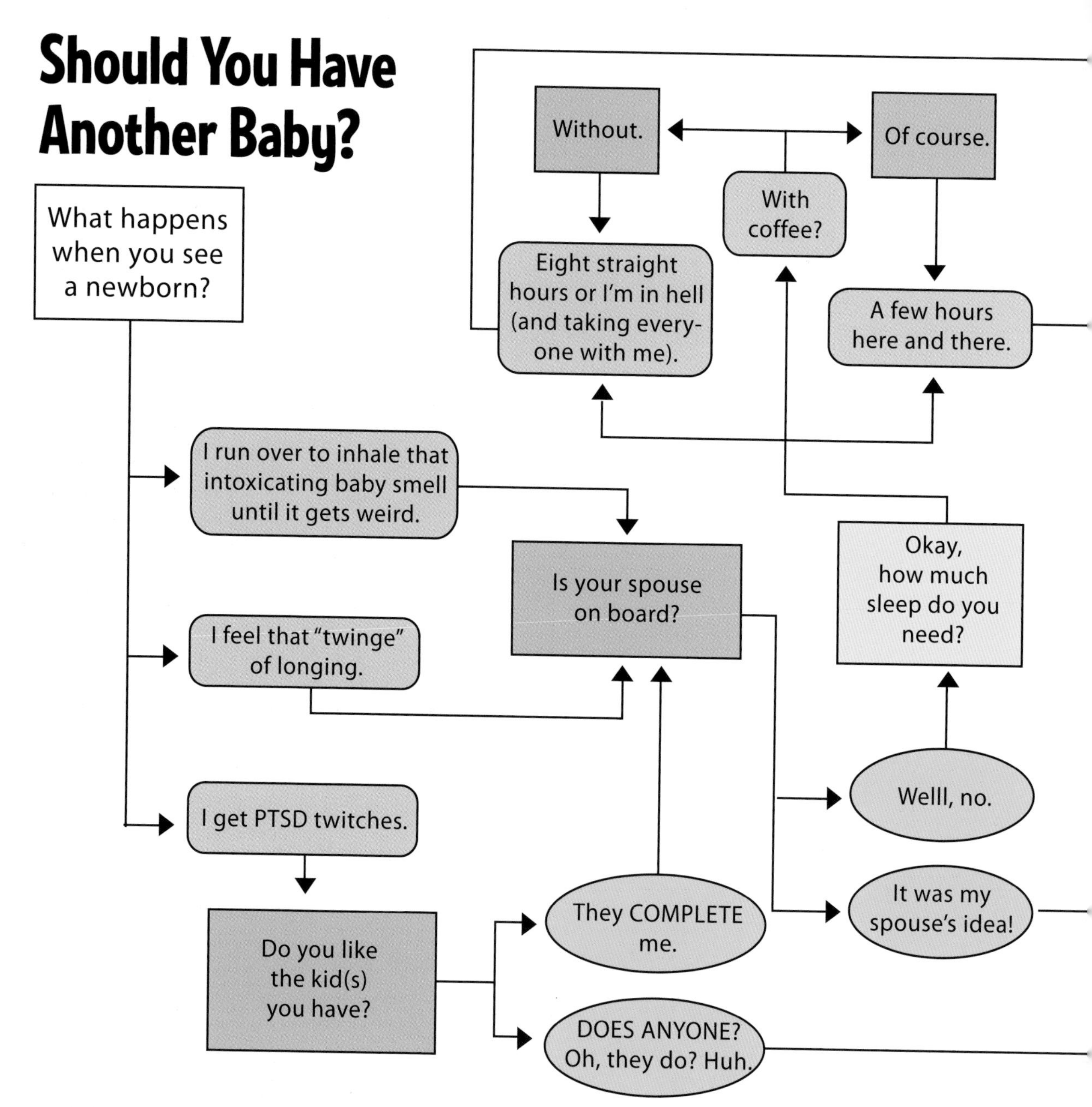

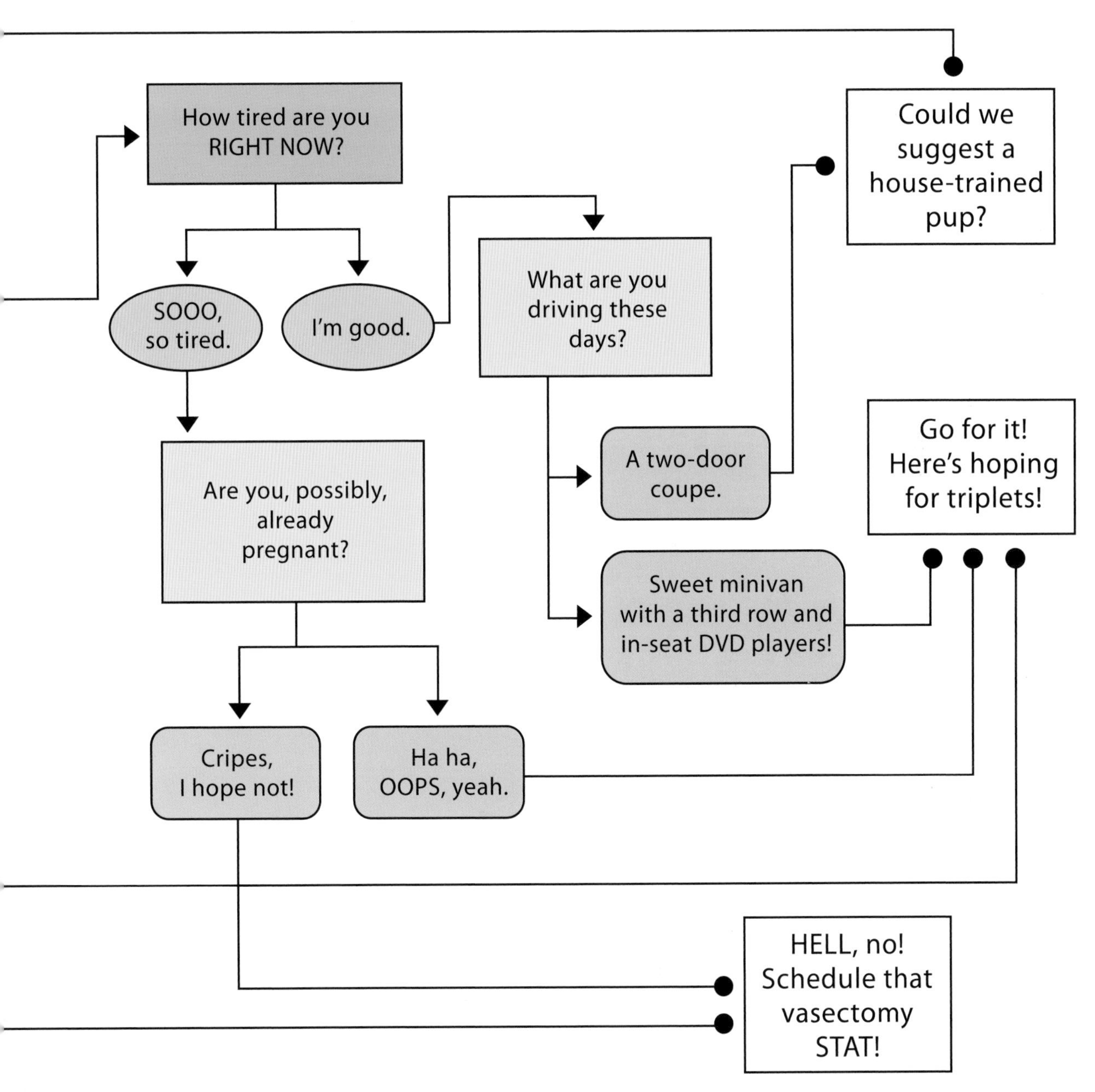
How tired are you RIGHT NOW?
SOOO, so tired.
I'm good.
What are you driving these days?
A two-door coupe.
Sweet minivan with a third row and in-seat DVD players!
Could we suggest a house-trained pup?
Go for it! Here's hoping for triplets!
Are you, possibly, already pregnant?
Cripes, I hope not!
Ha ha, OOPS, yeah.
HELL, no! Schedule that vasectomy STAT!

Post-Birth Conditions Your OB Might Forget to Mention (Don't Say We Didn't Warn You!)

In the interest of fully educated decision making, we've outlined a few post-birth conditions that might be missing from your pre-baby prep reading. Episiotomy shmesiotomy—here's what's really gonna hurt.

Postpartum Discharge Disorder

The paralyzing anxiety that hits both mom and dad as soon as they wave buh-bye to the nursing staff in the maternity unit and pass through the hospital's sliding glass doors and out into the parking lot to buckle their newborn into his car seat for the drive home. One or both parents may be given to repeating, "Why would anybody let us take a baby home?" while banging their heads against the car window.

Momsomnia

Also known as sleeping for the rest of your days with one ear on alert for sudden wails, coughs, and, in later years, the squeak of a bedroom window sliding open to allow a teen to slip out at night. Sleep deprivation generally starts the night you bring your baby home (or after the really good drugs wear off) . . . and continues until your child is making regular mortgage payments on her own. This condition affects ten out of ten mothers. Fathers are unaffected; the Y chromosome has a protective effect against this disorder.

Chromatic Coordination Syndrome

The compulsion to match a new baby's outfit to her hair ribbons, shoes, binkies, stroller blanket, and, in severe cases, the stroller itself. This obsession is considered a newcomer to the growing list of social (media) diseases and typically develops after binge-browsing Pinterest boards like "My Imaginary Well-Dressed Toddler." C.C.S. can often go undetected for weeks or months but becomes noticeable when a parent cannot bring herself to "Just throw on a onesie already, and let's go!"

Parental Separation Anxiety Attacks

Sudden parental separation anxiety (PSA) attacks often blindside new parents who sincerely *believe* they are desperate to leave the baby with a sitter and have some "grown-up time," but aren't really ready to let go. In the grip of such an attack, a parent (usually Mom) may sneak off to the restroom to check in with the babysitter every ten minutes (leading those around her to wonder if she has "a UTI or something"). In extreme cases, a parent having a PSA attack will grab anything in the vicinity that is remotely "baby-shaped"—a sack of flour, balled-up sweater, couch cushion—to create a faux "bundle" to rock in her arms.

Sensory Obliteration Syndrome

Known simply as S.O.S., this condition occurs when the body attempts to rebalance the heightened senses associated with pregnancy . . . and tips too far into the desensitization zone. S.O.S. often develops in tandem with Momsomnia and becomes noticeable when the baby is six to eight weeks old. By that time, moms are generally staggering around like zombies, unaware of much beyond the baby's need to be fed or changed. It's not uncommon for things like wicked BO or spit-up in the hair to go unnoticed for days.

Cranial Inhalation Compulsion

Going by innocuous names like "whiffing" or "huffing," cranial inhalation is on the rise among parents irresistibly drawn to the feelings of euphoria associated with sniffing babies. Baby aroma is highly addictive, and huffers have been known to lose interest in all other activities. Some will sit for hours, breathing in the intoxicating scent of freshly bathed babies while ignoring their Netflix queue and allowing their Facebook feed to atrophy. If you need help, dial 1-888-BABY-HUF or tweet @SniffMyBaby.

Acquired Distraction Disorder

Marked by an impatient *Now, what were we talking about?*, Acquired Distraction Disorder (A.D.D.) is the progressive loss of the ability to follow a train of thought. A.D.D. typically develops among parents with toddlers who've just learned how to run. The adult brain becomes overwhelmed with the strain of excessive multitasking and begins shutting down "nonessential" functions to conserve energy for chasing tiny humans intent on leaping from garden walls and licking electrical outlets. Fortunately, A.D.D. lasts only until middle school, when children stop interacting with their parents altogether.

Pregnancy Craving Timeline

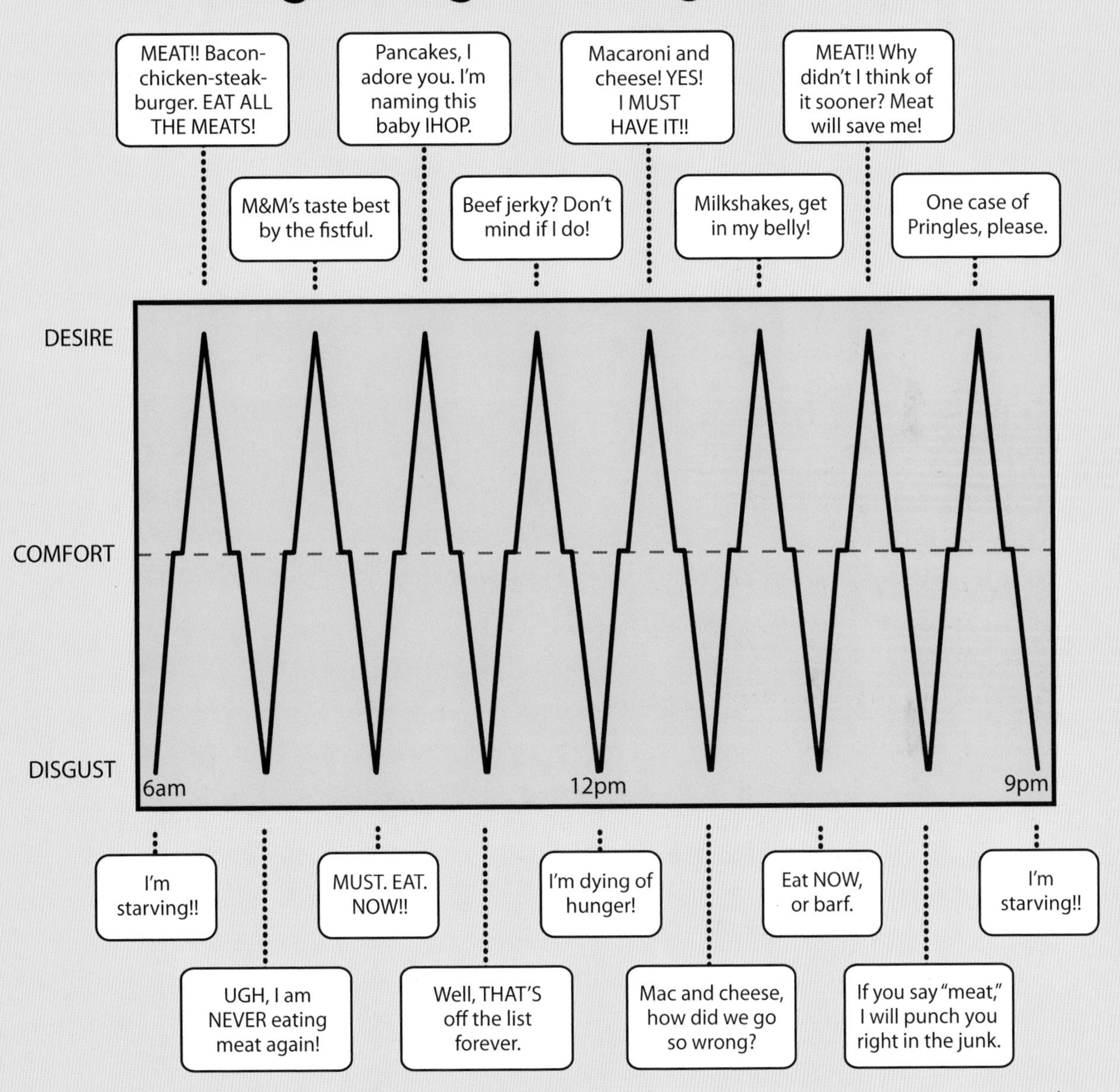

Mourning Sickness

The grief a pregnant woman
feels when
her favorite meal
winds up on the
things-that-make-me-barf list.
Forever.

Collective Pregnancy Unconscious

The Jungian state of mind
that connects every pregnant woman on the planet, allowing her
super-unique baby name
to become the
most popular name of the year.

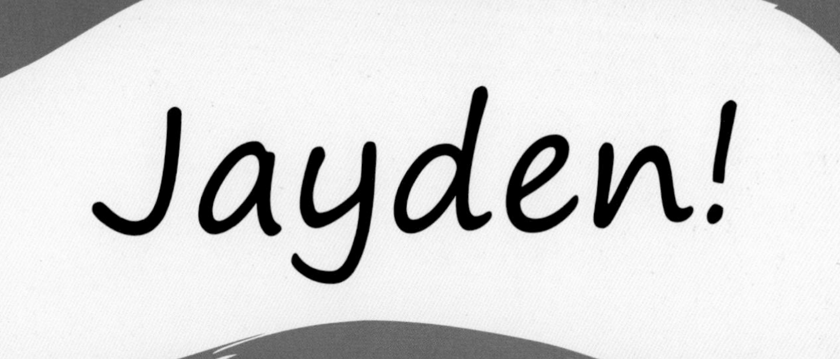
Jayden!

Anatomy of the Hospital Bag

Evolutionarily hardwired as "collectors" and "gatherers," women instinctively pack for EVERY POSSIBLE contingency. And while that typically entails schlepping around bags that weigh a metric crap-ton, one result of this excessive gathering is that Collector-Gatherers are rarely caught off guard by such situations as . . .

- A sudden rainstorm. They have umbrellas.
- A severe paper cut. Ditto bandages AND Neosporin.
- Chilly movie theaters. Of course they have a sweater. Do you want it?
- Plummeting blood sugar. Their snack stash rivals most hotel minibars.
- Lunchtime workout. With gym clothes, cross-trainers and shower kit in hand, they can be on the treadmill, then back at their desks within an hour.
- The impromptu dinner invitation. They carry heels and a string of pearls, for just such eventualities.

That's merely a small sample of what the average Collector-Gatherer totes around every day, a habit that has led some scientists to speculate that women must be distantly related to nature's other schleppers—the snail, hermit crab, and land tortoise. Indeed, that leading theory is thought to explain why women in labor typically check into the maternity ward with only slightly less gear than research scientists require for an expedition to McMurdo Station in Antarctica.

Let's examine what's typically hauled into labor-and-delivery in the Collector-Gatherer's hospital bag . . . and what actually gets used.

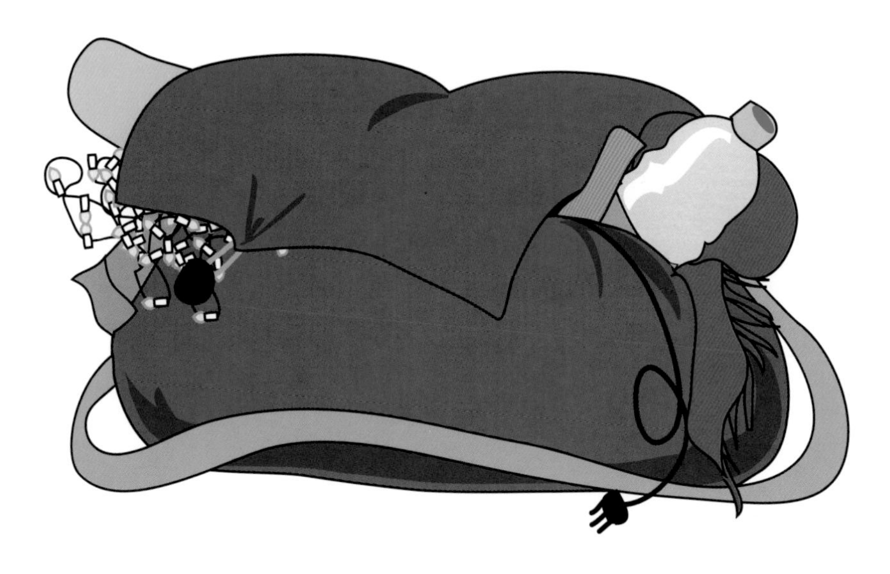

People
People
ple
NETFLIX
Baby's
1st Year
BIRTH
PLAN
LUNA
LUNA
LUNA

What Gets Brought

Fuzzy socks, most flattering silk robe, favorite couch throw (for when she's cold)

Oscillating fan (for when she's hot)

Personal pillow

String of decorative lights (Fluorescent hospital lighting? Oh no!)

Framed pictures of pets and some potted African violets (So homey!)

Playing cards

The latest *People*, *Us Weekly*, *InTouch*, *Star* and *OK!* magazines

Back massager, lotion and aromatherapy candles

iPad loaded with Netflix favorites and Push Playlist ("Ride of the Valkyries" and "Born to Run" queued)

Speakers for the iPad

Birth Plan binder

Hair dryer, curling/flat iron

Shampoo, conditioner, bath gel

Makeup (to look fresh and beautiful for those #BirthJustHappened pictures)

Skinny jeans to wear home

Assortment of going-home outfits for baby

Stocked diaper bag

Baby Book (It's never too early to start!)

Energy bars and lollipops

Bottled water

What Gets Used

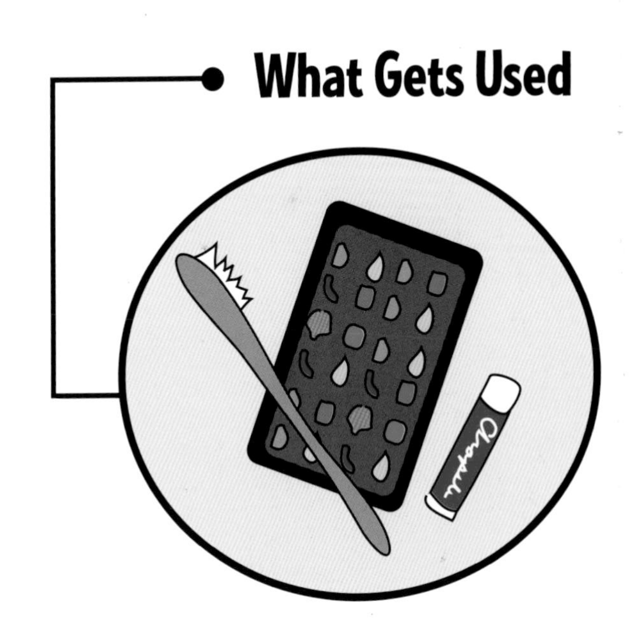

Maternal Classification System

It's human nature to sort stuff into categories. We just can't help ourselves. Sorting and categorizing is how we make sense of the world around us (and why stamp collecting remains a hobby when only Grandma sends letters by mail these days). But you know who *really* loves to sort and classify things even more than stamp collectors? Biologists! Recently, biologists turned their attention to a long-overlooked subspecies of *Homo sapiens*: *Mater Matris Americanus* . . . aka the American Mom.

American Mom was previously thought to be a single homogenous subspecies. Then Working Mom was discovered. Followed by Soccer Mom. And finally Tiger Mom. As biologists dug deeper, they realized there were still more variants that had never before been catalogued. While scientists debate just how many there are, it's generally accepted that *Mater Matris Americanus* comprises these four primary groups:

Hands-On Mom (*Mater Matris Americanus Helicopteris*)

Named for the iron grip she maintains on her offspring and her ability to hover tirelessly,

Hands-On Mom ensures that she remains involved with every aspect of her offspring's lives even as they morph into adulthood. Included in this group:

Tiger Mom: This ultra-strict species indigenous to China—known to researchers as *M. Chua Helicopteris*—is said to keep a stronger hold on its offspring than *Americanus Helicopteris* and maintain even higher standards of excellence and achievement. As Tiger Mom steadily migrates throughout the continental United States, biologists are watching to see if she completely crowds out *Americanus Helicopteris* or if crossbreeding leads to a still more demanding super-species.

Team Mom: Soccer Moms are part of this migratory family, which travels in minivan formation and is found primarily on sports fields and indoor courts, foraging for goldfish crackers, juice boxes, and passably clean uniforms.

Backstage Mom: Known as Gypsy Rose Mom in New York and California, her natural habitats include theatre wings, pageant circuits, dance studios, reality TV show sets, and anywhere else she can boast, "My kid's gonna be a star!"

Perfection Mom
(*Mater Matris Americanus Perfectus*)

Perfection Mom—also called Do It All Mom in the Southwest and I Don't Know How She

Does It All Mom in the Northeast—shares many of the same characteristics as Hands-On Mom. The primary distinction is their hours of activity: Hands-On Mom is more active during the day, while Perfection Mom is nocturnal, preferring to stalk Pinterest boards late into the night. Perfection Mom also has a pronounced compulsive streak and a shrill distress call. Under pressure, she can often be heard shrieking, "Anything worth doing is worth making everyone batshit crazy till it's done RIGHT!"

Outsourcer Mom
(*Mater Matris Americanus Absentia*)

Outsourcer Mom (sometimes called Working Mom) is the "worker bee" of the *Mater Matris Americanus* species, known for the long hours she spends juggling the numerous tasks that keep her community humming along. When it comes to reproduction, the Outsourcer's primary responsibility is gestation. After birth, she typically turns her offspring over to the elders of the community—variously called Papa, Papi, Grandpa, Granddad, Grandma, Grammy, or MeeMaw, depending on the region—or to a group of apprentice moms within the collective called Daycare.

Know-It-All Mom
(*Mater Matris Americanus Omniscience*)

The popular name for *Americanus Omniscience* derives from

the Know-It-All's voracious appetite for information. Each day, she consumes three times her body weight in Internet news and remembers every fact she's ever read, so she really does appear to "know it all." Yet, despite her Wikipedic stores of information, Know-It-Alls are generally regarded as the pests of the Mom world for their continuous tsk-tsking, "I'd never let my child do/eat/play/ride/watch ______."

The Smug Foods Mom is also found in this group, though she is primarily drawn to parents who don't grind their own wheat into flour, serve milk straight from the cow, or are unwilling to debate the ramifications of GMOs and European food standards.

Inverse Dilation Rule

The level of bravado for
white-knuckling it through
natural childbirth
is inversely proportionate
to the time it takes to
demand an epidural
—STAT!—
when labor "gets real."

DARWIN'S PARENTS: ADAPT OR DIE!

Whoever coined the phrase "Change is good" clearly never woke up every two hours to feed a newborn. Or paced 26.2 miles round the living room, trying to soothe a screaming infant. Or went ten days without showering because she was too exhausted to care. Or notice.

Yep. Change sucks more than a big ol' black hole. Which undoubtedly began to dawn on you once you brought that little bundle of joy home, and all those "necessities," like sleep, food, sex, a hot shower, and functional hearing, suddenly felt like luxuries you'd never have again.

Suck it up, Buttercup. Eventually, you will sleep more than two hours at a stretch. And change clothes. And grab a shower. You may even have sex again. But just when you think, *It's smooth sailing now!*, the kid will stop napping or start potty training or develop an irrational fear of Mickey Mouse right after you arrive at Disney for a week-long vacation. And your life will go to hell once more.

Change: It's the new black. If only it hid spit-up stains.

Rorschach Burp Cloth Test

The use of
spit-up stain patterns
to evaluate a
sleep-deprived Mom's
psychological
state.

Will You Shower Today?

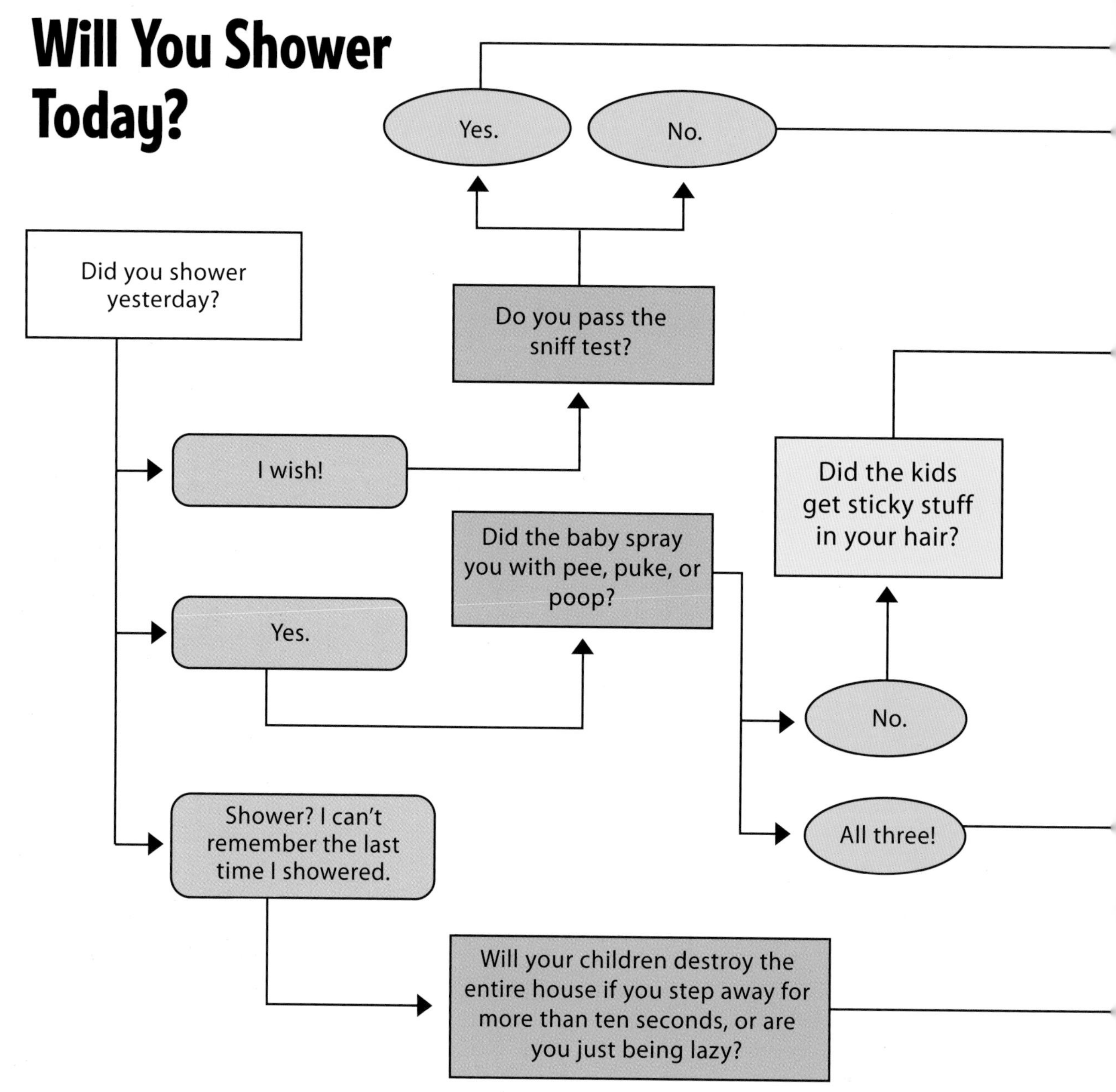

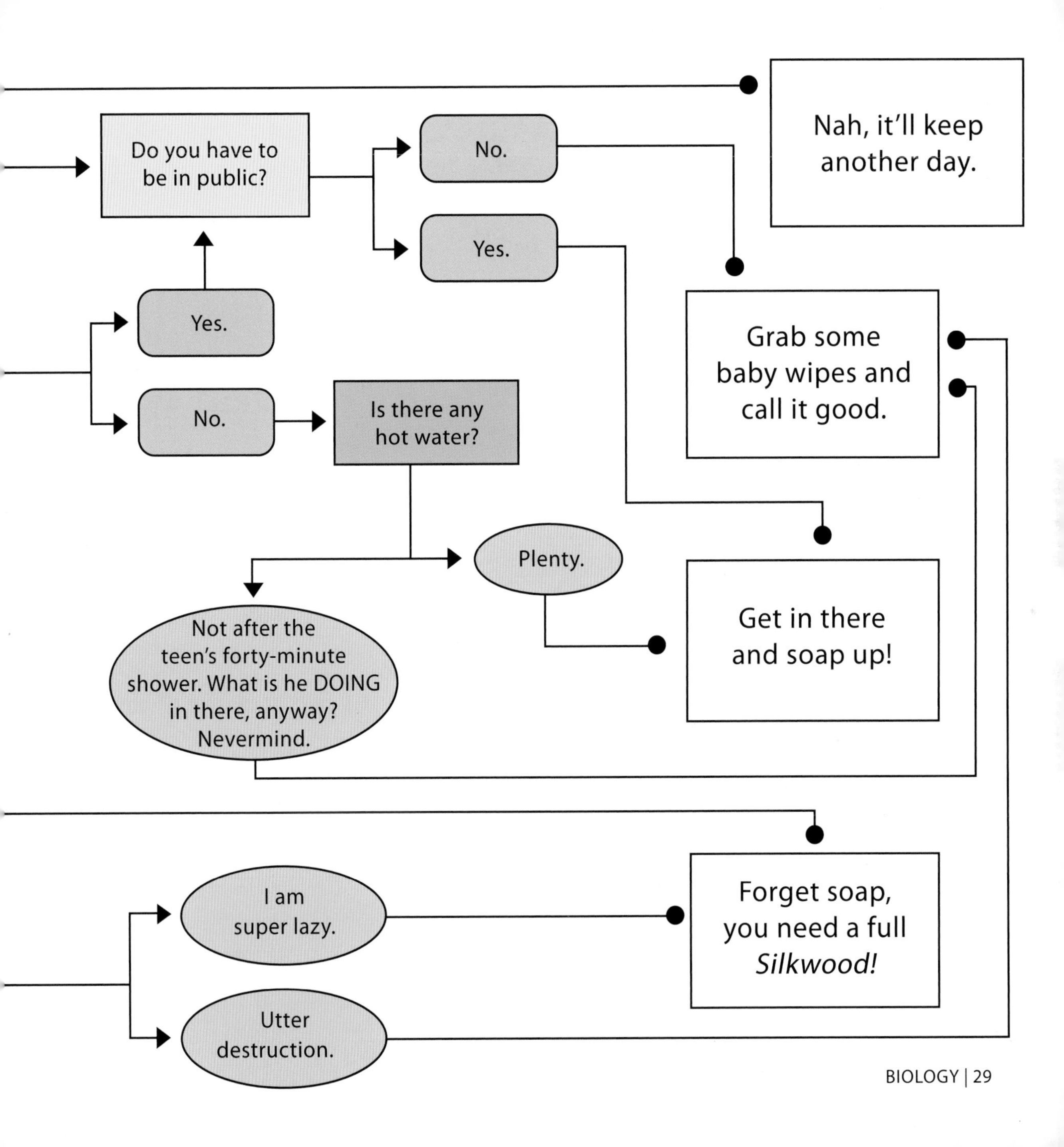
Nah, it'll keep another day.
Do you have to be in public?
No.
Yes.
Grab some baby wipes and call it good.
Yes.
No.
Is there any hot water?
Plenty.
Get in there and soap up!
Not after the teen's forty-minute shower. What is he DOING in there, anyway? Nevermind.
I am super lazy.
Forget soap, you need a full *Silkwood!*
Utter destruction.

Steady State

The most eagerly
anticipated moment
of every exhausted new parent's life:
the night that,
no matter how many times
Mom peeks in,
the baby sleeps straight
through till morning.

Natural Selection Matrix

Children will naturally choose the parent or supervising adult who best meets their "needs" at any given moment.

Paleosexuality: Evolution of Mom's Sex Drive

The complexities of the maternal sex drive and its characteristic traits have mystified paleosexologists (not to mention eager-beaver spouses) since the first *Homo habilis* wanted to get busy and his mate suggested he acquaint himself with his opposable thumb instead. Recently, however, teams of researchers have finally been able to piece together the five key phases of Modern Mom's sex drive as she adapts to her ever-changing environment.

A. Not-Yet-A-Mom

Roughly spanning the years between The Third Date and Before Kids, this period (aka the Honeymoon Era) is marked by spontaneous bursts of "God, I want you NOW!" and keen interest in the *Kama Sutra*, *Magic Mike,* and multiple shades of grey.

Tools & Artifacts: Handcuffs, massage oil, "clothing" marked with a large pink VS.

B. Wanna-Be Mom

Coinciding with the Cohabitation/Marriage Era, in this phase of development we see a substantial decrease in spontaneity and a corresponding increase in highly scheduled, results-driven activity. Language shifts significantly among wanna-be moms, as purring "Oh yeah, let's do it," becomes "I'm ovulating! LET'S DO IT! NOWNOWNOW!"

Tools & Artifacts: Ovulation kit, iCal marked "TONIGHT!" at thirty-day intervals.

C. You Have GOT to Be Kidding Me! Mom

The first of several distinct stages in the Post-Baby Epoch, this period (aka The Ice Age) is identified by a heavy freeze on all activity until the new mom is certain it won't "hurt like a mofo." Attempts to thaw relations prematurely, because "C'mon babe, I haven't had sex in ages!" are typically met with "Are you kidding me?!?" and an icy suggestion to "go whittle your spear." Paleosexologists differ on duration of the Ice Age. Some maintain it persists just a few months while others insist it can last till the "baby" finishes medical school. Interestingly, the freeze appears to thaw faster in regions where dads take on fifty percent or more of baby-care duties without being nagged.

Tools & Artifacts: Premium personal lubricant for him, YouPorn.com bookmarks.

D. Exhausted Mom

The middle era of the Post-Baby Epoch is the first time a new mom might dimly recall that she actually likes sex; she just can't keep her eyes open long enough to participate. The only thing that turns her on now is the thought of eight to ten consecutive, uninterrupted hours of sleep.

Tools & Artifacts: Earplugs, sleep mask.

E. Digital Mom

The twin discoveries of the DVR and the ability to fill a Netflix queue with kiddie cartoons marks a fundamental shift in mom's sex drive, as she suddenly finds herself with a full twenty-four minutes at a time in which small humans are not hanging off her or screaming in her face. Some paleosexologists believe this is where siblings originate.

Tools & Artifacts: TV or tablet, Video On Demand subscription, bedroom door lock.

Mach's Date-Night Principle

A baby will exert
a gravitational pull on a
new Mom so intense
that despite numerous text messages
assuring her that the baby is,
in fact, still breathing,
she will need to
race home
to check for herself.

The Modern Parent's Dinner Devolution

The North American Parent is an omnivorous eater, consuming a wide variety of foods to ensure adequate nutrition. Historically, the Traditional North American Parent evolved from eating a steady diet of pizza, Cap'n Crunch, and beef jerky during the College Era to adopt a healthier, more balanced diet in the Double Income–No Kids Epoch.

Modern parents, having survived a cataclysmic shift in lifestyle as a result of The Great Birth Event, may start out with the best of nutritional intentions but find that their meal planning quickly devolves into *Hunger Games*–style chaos as they chauffeur their brood from school to Kumon to baseball to gymnastics to scouts to piano to . . . *Shit! I'm starving!*

Stage 1

Nutritious, home-cooked meal made with ethically raised locavore meats, sustainable whole grains, and organic produce bought daily at the farmers' market.

Stage 2

Organic roast chicken and assorted sides picked up at Whole Foods after Mommy & Me.

Stage 3

Chicken dinner picked up from Boston Market on the way home from daycare.

Stage 4

Bucket of KFC, hastily grabbed at the drive-thru after the Little League game goes into extra innings.

Stage 5

Delivery pizza wolfed down an hour past bedtime because *someone* forgot to order until 7 p.m.

Stage 6

Cold cereal eaten standing over the sink while checking homework and packing the next day's lunches.

Stage 7

The uneaten mac and cheese scraped from the kids' dinner plates.

Stage 8

M&M's scarfed from the minivan cup holder while flooring it from the office to aftercare before those damn $1-per-minute late charges kick in.

Stage 9

Stale pretzel stix, string cheese, and stray raisins scrounged from the bottom of a kid's backpack because you've been stuck on shuttle duty since 7 a.m.

Proof by Exhaustion

When learning a new skill,
a child will continue
to climb, scale, and traverse
until her parent is
utterly spent.

TOXIC WASTE

Early childhood experts love to say that "every child is different," that "one child's experience is not guaranteed to be another's."

To that, we say, "Bullshit." *Our* research indicates that as wholly unpredictable as parenthood is, there are three things you can absolutely count on once you throw a kid (or several) into the mix:

- A vomiting child will nail you with the vomit, especially if you're late for work.
- The kid with the nastiest green snot running from her nose will find a way to wipe it on your shirt.
- An empty diaper caddy guarantees a diarrhea blowout.

Our irrefutable—and (unfortunately) reproducible—findings clearly demonstrate that kids are highly enterprising little mess-makers. Bless their hearts and sticky little hands, kiddos are always so eager to share, blissfully unaware of just how disgusting they actually are. While you're busy bumping a level-four biohazard suit to the top of your Amazon wish list, they're engrossed in rolling pellets of poop on the carpet like marbles. Just count your lucky stars they're not smearing that shit on the walls or . . .

OH, CRAP! Too late.

Proof by Contradiction

The "I Call Bullshit" moment
when you discover
that your kid's playmate,
whose mother SWORE
he was potty-trained,
is not, in fact, potty-trained.

AT ALL.

When Will Your Kid Puke?

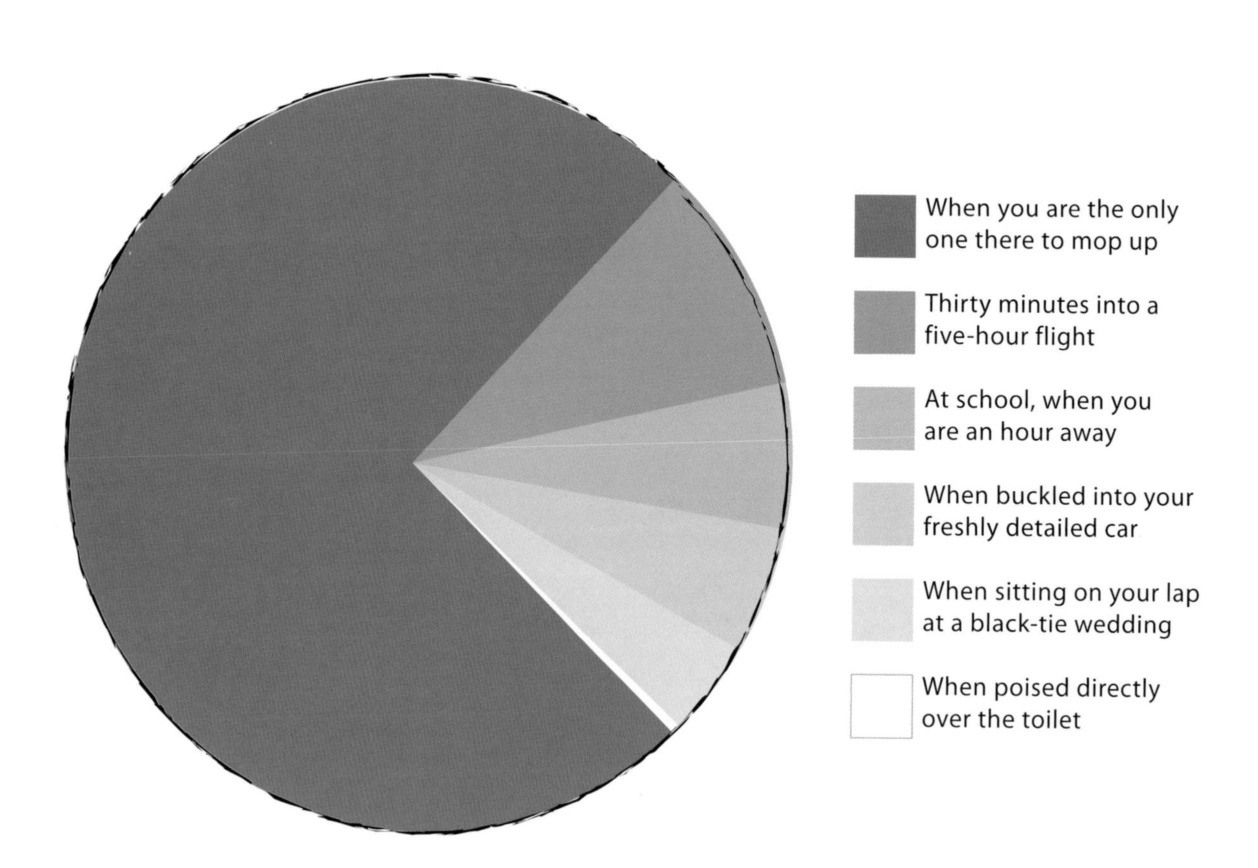

Field Guide to Childhood Ailments

Kids are determined to put their mouths on everything—the escalator handrail at JFK International Airport; the toys that kid with hand-foot-and-mouth disease was playing with at the pediatrician's. They are perfect petri dishes, continuously incubating whichever nasty germs are floating around your neighborhood and generously doing their part to ensure that everyone shares their sniffly, sneezy, achy, whiny misery. Being stuck at home is never more fun than when you *and* your child are sick together. Unless it's when your tot is hopped up on ibuprofen while you are praying that death takes you swiftly.

Children under five average nine to twelve colds a year. Meaning, your kid's nose will ooze yellow-green slime until he or she enters elementary school. But between the expected run of coughs, colds, and flus, you can count on your tiny Patient Zero to come down with a few of these other highly common conditions too.

Street Sleeping Sickness

Narcoleptic episodes that occur whenever a child is buckled into a vehicle. It's thought that the "click" of the car-seat buckle and the start of the motor work in tandem to produce a potent sedative that knocks out even the most energetic child. Attempting to remove the child from the vehicle, however, will rouse him like a shot of adrenaline, and he'll be bouncing off the walls till well after midnight.

Irritable Appetite Syndrome

A stubborn stomach disorder that manifests with a refusal to eat a favorite food prepared exactly the same way it's been prepared the last hundred times it was made. Other indicators include *I hate that now!* and *It tastes yucky!* May resolve with a double dose of *Fine. Go to bed hungry.* Or not.

Ferris Fever

A general feeling of sickness that primarily affects children elementary school age through high school and may include a mix of any of the following: headache, dizziness, nausea, very low-grade fever, vague stomach pain, and overall lethargy. Symptoms typically develop within moments of waking up and realizing it's a school day. Often epidemic when *Ferris Bueller's Day Off* pops up on cable. Parents frequently suffer similar symptoms, especially if they grew up in the '80s.

Mess-Related Myopia

Blurred vision that occurs whenever the words *Clean up your room* or *Pick up your toys* are uttered. Symptoms include moaning, *It IS clean!* together with an inability to find anything in the "clean" room. Vision usually remains impaired until the kid leaves for college and that room is turned into an office.

Related: Mother-Induced Myopia: Blind spots in otherwise normal vision that appear whenever Mom asks a child to "Pause the TV (or video game)," to bring her something from another room. Symptoms: Repetition of *I can't find it. I don't see it. It's not there.*

Acquired Band Aid Boo-Boos

The sheer number of bandages covering the body can make it seem as if a child has toddled through an abattoir. However, this skin condition is actually mild, noncontagious and most often triggered by a new box of Strawberry Shortcake or SpongeBob Band Aids. Kids typically outgrow A.B.A.B.B. by age five or six. Flare-ups resembling idiosyncratic Chinese kanji may appear on backs, necks, arms, and legs, starting in the late teen years.

Peekaboo Fever

The high fever that may accompany any illness in children three and under. This fever generally spikes in the middle of the night, causing severe parental heart palpitations, particularly when the only option for medical attention at that hour is the community hospital emergency room, filled with gunshot victims and MRSA bacteria. It's known as the "peekaboo" fever for its tendency to drop back to normal range after a parent makes a panicked call to the pediatrician, demanding a house call. The bigger the parental fuss, the lower the kid's temperature will be in the morning when the doctor finally sees her.

Parental Frequency Hearing Loss

A progressive disorder, beginning in early elementary school, in which children lose the ability to hear anything a parent is saying. Hearing loss is generally limited to the parents' voice frequency, though it can worsen to include teachers' voices as well. Parents may find that a tenfold increase in volume and near-constant repetition may help compensate. Sudden reversals in hearing loss, triggered by electronic device pings, have been known to occur.

Chutes & Madder Syndrome

Severe psychic distress resulting from the loss of any competitive game. Symptoms include shrieking, stomping, and throwing of game pieces. Distress can increase dramatically if the child feels that he's being "allowed to win," or conversely, that he "always loses." Distress may also occur if the child feels the game is "not fair" or the opponent is "too big." The end result: No one can win. Ever.

When Your Kid Will Poop

Carseat Carbon Dating

Advanced technology
used to determine
how old the foodstuff
collected
under the babyseat
actually is.

Toxic Shock Syndrome

A parent's psychological state
on discovering that her tot
has gotten his diaper off
and smeared the
walls, crib, bedding, and himself
with poop.

Poopology 101: The Gushy, Gassy, and Gooey

You don't need to be a PhD biologist to figure out that kid poops come in a wide variety of shapes, sizes, odors, and consistencies. (If you've encountered the mustard ooze that follows a round of antibiotics, you know *exactly* what we're talking about.)

Still, PhD or no, it's easy to get faked out. You might catch a whiff of something that if distilled correctly could be weaponized, then steel yourself as you open the diaper and find . . . nothing. Just a bit of (debatably) harmless gas. Expecting sewer sludge and getting gas is what we in the field call a "happy accident." Especially if you're low on wipes at the moment of expulsion. (Pro Tip: Don't get low on wipes. Ever.)

But there will be other times when you anticipate a tiny toot and get decidedly more. Those are the moments that keep scatologists in grant money. But since few outside the scatology field truly revel in examining the nuances of fecal waste, we've used proprietary computer modeling to forecast a range of "mixed consistency" scenarios that, during the diaper and potty-training years, will warrant an immediate detour to the dry cleaner or the auto detailer or a call to an industrial waste-removal team.

Vapor-Solid:

Gas accompanied by unanticipated solid waste, aka shart. Likelihood of occurrence:

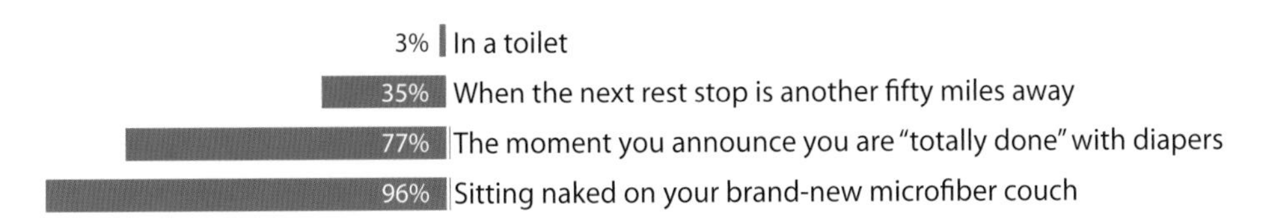

Solid-Liquid:

Runny, watery poop, aka diarrhea. Likelihood of occurrence:

0% After eating nothing but buttered noodles for three weeks

68% Twelve hours after playing at an indoor playscape

89% After unsupervised access to dried apricots

99.9% After Taco Bell

Vapor-Liquid-Solid:

The excretory trifecta—explosive, resounding diarrhea. Likelihood of occurrence:

54% After an apple-juice binge

78% Weekly during your three years of preschool

86% In the middle of Mommy & Me swim lessons

96% Ten minutes into *The Nutcracker* performance you bought $150 tickets for and now, thanks to an ill-timed apple-juice splurge, will miss entirely

DO THESE GENES MAKE ME LOOK CRAZY?

It's been said that insanity is doing the same thing over and over and expecting different results. If that sounds like you, fetching the toy your tot throws from his high chair again and again . . . and again . . . and again . . . then yes, you are going insane. And you can thank your little one for driving you there.

Still, our methodical research shows that you may relax and enjoy the ride once you realize that parenthood is *designed* to drive you completely crazy. (Tapping into that Xanax we urge you to stockpile will help a bunch too.) There's the boredom of coloring page upon page of Dora characters. The monotony of hearing *The Wheels on the Bus* repeated ad infinitum. The day-in-day-out-ness of putting food in one end and wiping it up as it comes out the other. There's the anxiety that you're doing too much, or not doing enough, or that you *are* doing enough, just not enough of the *right* stuff. And the guilt and doubt that only other parents in your kids' cohort can instill. (Not you, of course. The *other* other parents.)

And let's not forget the paralyzing shock when you hear your *mother's* words coming out of *your* mouth.

(But really, how else would you know you're doing it right?)

Biological Comparative

Secretly assessing
all the other babies
in your weekly playgroup
and concluding that your baby
deserves a
genius grant.

2
+3
6
+8
B
G

An Important Update to the DSM-V

Given the intense debate surrounding the conditions that were included and excluded from the *Fifth Edition of the Diagnostic and Statistical Manual of Mental Disorders* (the DSM-V or the "psychiatrists' bible") when it was published in 2013, we were not a bit surprised that the editors of the DSM-V decided to issue the following addendum regarding parents' mental health on the down-low.

Although this addendum—"Insanity: No, You're Not Crazy, You DO Get It from Your Children"—will be included in future printings of the DSM-V*, we received permission from the American Psychiatric Association** to reprint it here in an effort to raise awareness about the special conditions that may affect people once they have kids.

These conditions can strike healthy, rational adults at any time following the birth of a baby. And they have been known to last up until the time parents move that kid into his or her dorm room at college (hopefully in a state far, far away). While symptoms of these disorders are often similar (most notably repetition of the phrase, "Why? Why? WHY?"), physicians should be aware of their key differences to make accurate diagnoses.

---

* Not really.

** Like they'd talk to us.

Momochondria

The intense fear some mothers experience that the child they nurtured protectively in their womb for forty weeks will suddenly be struck with some incurable disease through routine contact with the outside world.

Symptoms: Persistent belief that every pimple is MRSA; every sniffle, SARS; every fever, Ebola. Excessive use of hand sanitizer—slathered over child's body. In extreme cases, Mom may insist that the child be contained in a "plastic bubble" through law school.

Delusions of Launder

The perpetual belief that one day, eventually, the laundry will get "finished."

Symptoms: Moms laboring under this delusion may initially appear upbeat, even enthusiastic, aiming to dominate the heaps of dirty clothes and pee-soaked toddler bedding. But as the laundry piles grow, these moms can sink into a depression as they ruminate on existential questions such as *Where does all this laundry come from?* and *Why is all this laundry here?* Fixating on "finishing" the laundry may lead to secondary physical problems, such as carpal tunnel syndrome, the result of folding endless pairs of teeny-tiny socks and superhero underwear.

PVSD (aka Persistent Volunteer Stress Disorder)

Acute mental trauma brought on by excessive stepping-up, volunteering, and/or otherwise "helping out" in the classroom and the PTO or for the sports teams, scout troops, dance recitals, Sunday School, or any other children's activity that parents are "invited" to participate in.*

Symptoms: Obsessive attachment to clipboard; rocking in a corner, repeatedly mumbling, "I'll do it. Count on me"; may experience flashbacks of contentious SAC or PTO meetings; often found to be hoarding vast caches of poster board for bake sales and car washes.

Anime Anxiety

Emotional distress brought on by overexposure to Japanese anime via Pokémon, Beyblades, and Yu-Gi-Oh! trading card games, YouTube videos, and television shows.

Symptoms: Panic attacks are often triggered by nonstop chatter (about the cartoon characters' health points, epic battles, powers, and damage done) and/or realizing just how much these plastic tops, figures, and trading cards actually cost. (See Neumann R. *et al.* in the *New England Journal of Medicine* for the case study of the mom who suffered a near-fatal case of sticker shock while browsing eBay for "super rare" Pokémon cards.)

Sideliner's Syndrome

The delusion that routinely watching every game/match/tournament/playoff of a given sport affords a parent infinitely more game knowledge than the person employed as team coach and/or referee. This leads to the compulsion to share that knowledge, loudly, from the sidelines.

Symptoms: Typically calm in other venues, parents with Sideliner's tend to become enraged and exhibit increasingly aggressive behavior when on or near the "sidelines" of a

---

* The DSM-V editors were divided on whether "coerced" was the more accurate term. Readers should be aware that "invited" is being used in the Mafia or gang members sense—meaning they're being polite, but refusal isn't an option.

sports field or court, on weekends or after school. These rages are usually triggered by a self-perceived "bonehead" move or call made by the coach or referee and preceded by high levels of anxiety that whatever "dumbass" is running this U6 team is ruining their kid's shot at a college scholarship.

Snackorexia

The persistent desire to only consume foods packaged in paper or plastic bags and/or "go cups" that fit in a car cup-holder.

Symptoms: Weight gain; mood swings; acne breakouts not seen since puberty; minivan littered with empty snack packages and crumbs ground into the carpet. Symptoms are typically preceded by a full season of travel sports. Snackorexia tends to worsen with miles and days traveled and the number of turnpike rest-stop meals consumed.

Lego Traumatic Distress Syndrome

A Lego-specific panic attack that overwhelms the parent who is "helping" his or her child complete a complex Lego model after realizing that a piece won't fit on Step 587 because of a mistake they made back on Step 28. Followed by the immediate realization that the whole model must be disassembled and started again.

Symptoms: Nausea, dizziness, shortness of breath, heart palpitations. Frequently followed by an uncontrollable urge to smash the entire model to the ground, beat the pieces with a sledgehammer, douse the entire thing in gasoline, then toss a lit match onto the pile, while dropping a payload of F-bombs.

Chronic Festivity Fatigue

The extreme exhaustion that results from marking every holiday with a blowout celebration for the delight of children. This is a new, fairly controversial diagnosis in the DSM-V. Previously, parents with symptoms of Chronic Festivity Fatigue (C.F.F.) were written off as "party poopers." However, new research suggests that parents with C.F.F. are simply worn out by the effort of making every aspect of their lives "Pin worthy."

Symptoms: C.F.F. often starts out as casual Pinterest browsing. But among those carrying the DIY-1 or DIY-2 genetic mutations for extreme crafting, casual browsing may lead to binge-shopping for knick-knacks to redecorate the house for every holiday—including the ones no one gives a shit about, like Flag Day. Progressive C.F.F. often leads to what experts call "seasonal blurriness," in which sufferers allow decor from one holiday to "bleed" into another, keeping Christmas lights lit through June or eating Peeps out of season. Untreated, C.F.F. develops into a type of holiday catatonia, in which a person loses all interest in hiding Easter eggs or giving out Halloween candy. For more in-depth reading, see Kringle, K. in the *Archives of General Psychiatry*, a case study of the mom who super-glued her Elf to a shelf, ranting, "I'm sick of this little creepy sonofabitch. He's not moving again. EVER!"

Central Nervous System

The modern parent's central nervous system integrates information it receives from a host of outside sources—the surrounding environment, CNN, those Mommy & Me bitches—to coordinate and/or influence the activity of all parts of the body. Functions of the brain previously devoted to speech, movement, planning, and memory are now remapped to target the new primary function: keeping little thrill-seekers bent on self-destruction alive.

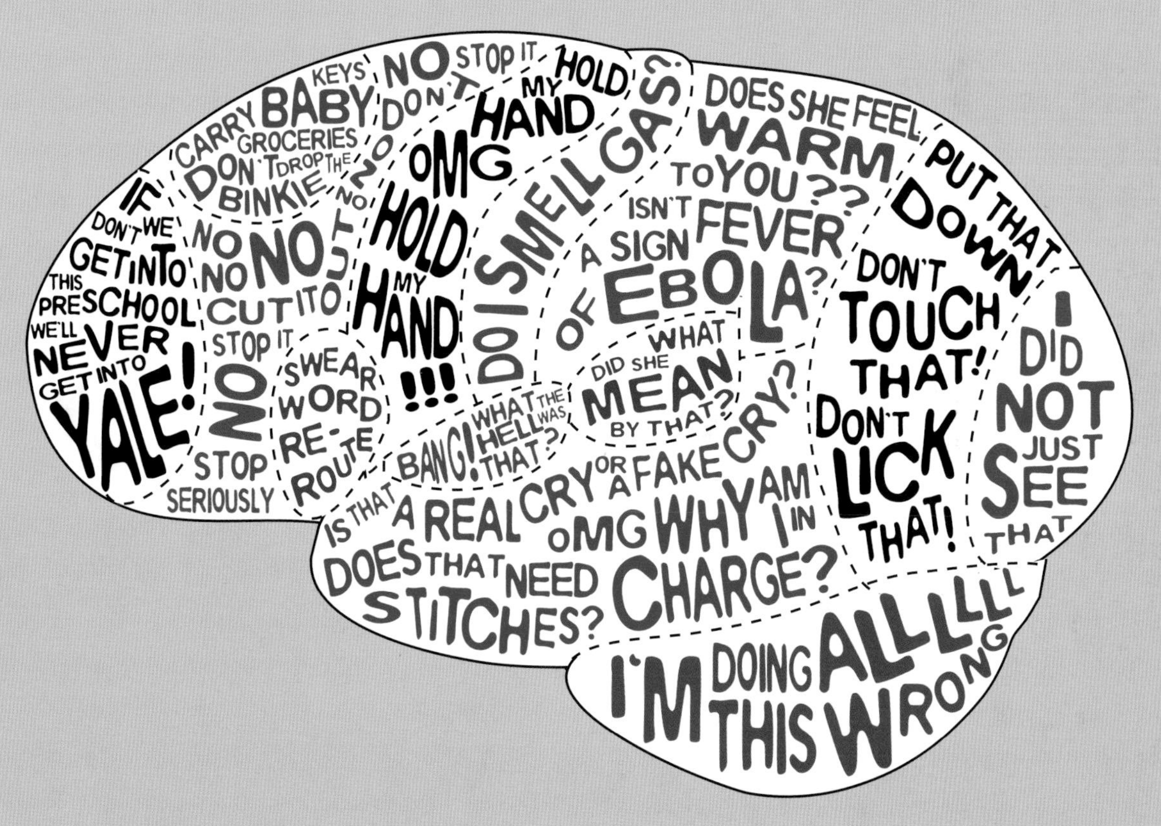

Passing of the Noble Gases

The time-honored tradition of
handing down
socially unacceptable
behaviors
from generation to generation.

Mendel's Principle of Finger Pointing

Fruit flies and your pink-cheeked angel. Biologically speaking, what could they *possibly* have in common?

Apart from a shared interest in fruit and a propensity for being little pests, the two are textbook examples of how traits are passed from generation to generation, using the simple concept those fruit flies demonstrated back in Biology 101: A + B = mashup of AB.

Of course, when it came time to do your own crossbreeding, you hoped your little darling would inherit only the very best from you and your partner—Mom's blue eyes; Dad's thick dark curls—while the worst traits would get buried deep in the attic of the double helix, never to be seen again . . . or at least not in the generation under your care.

But when you roll the genetic dice, sometimes you win, and sometimes you just want to bang your head against the wall. Science tells us that the most annoying traits always come from your partner's side of the family.* And since "credit" is the currency of scientific discovery, we can thank the German monk Gregor Mendel, the father of modern genetics, for helping us spin the positive traits in *our* favor . . . while pinning the blame for those less desirable traits squarely on our partners.

From You	From Your Partner
Creative Problem Solver	Sly Manipulator
Savvy Negotiator	Smart-Ass Back-Talker
Decisive Leader	Bossy Dictator
Focused Perfectionist	Constant Nitpicker
Energetic Dynamo	Hyperactive Fidgeter
Inspired Storyteller	Bald-Faced Liar
Skilled Organizer	Compulsive Neatnik
Shrewd Entrepreneur	Greedy Bastard
Suave Charmer	Perpetual Ass-Kisser

*Not always, but it feels good to point the finger, doesn't it?

Pavlov's Highchair

How your baby "conditions" you

to immediately pick up

the toy he's thrown to the floor

and give it back

each and every time

he throws it.

"Chemistry is not just when it stinks and bangs."

- Anonymous

Chemistry

CHEMICAL REACTIONS

Once you've had your kids, you're going to want to leave the house, if only to get a break from the On Demand preschool TV shows your tot will subject you to, twenty-four/seven. Get a few episodes into *Peppa Pig*. You'll see.

The best part about venturing out with your kids, maybe to some indoor playscape where you can station yourself at the single exit to prevent anyone from smuggling your kid out, is that you'll meet other parents—maybe parents who share the same level of anxiety you do. Unfortunately, the worst thing about venturing out with your kids is that you'll meet *other* parents who are certifiably looney tunes.

Making friends with other parents who would have you over to watch preschool TV at *their* house (perhaps even with some hummus, a little baba ghanoush, and some adult beverages) involves balancing more chemistry than an eHarmony questionnaire. So many variables: Can the kids get along for an hour or two without hitting or biting? Do the moms like each other? Can the dads at least find some common ground around football or the offsides rule in soccer?

If you're lucky, you have the kind of chemistry where everyone is so simpatico, you start doing stuff together like renting a beach house or starting a blog.

And if you're not so lucky? Well, have you heard of the Hindenberg?

Mendeleev's Table of Parental Elements

It's generally accepted that it was Russia's Dmitri Mendeleev who organized the original periodic table of elements. Except in Germany, where they're all, "Our guy, Julius Lothar Meyer, made an element table too! He's the father of OUR periodic table. So there!" Such whiners. Still, in Mendeleev's defense, he did grow up in Siberia in the mid-nineteenth century, so it's not like he had anything better to do. Poor guy, he didn't even have Pong.

So, yes, chemistry students everywhere but Germany have Mendeleev to thank for that ubiquitous chart hanging on the wall of classroom chem labs. But what's less commonly known is that back when Mendeleev was president of his neighborhood homeowner's association, he also organized an elemental table of families, so that parents could better identify which moms and dads were just a few electrons short of a full valence. Mendeleev won the Folderal Prize for his efforts, providing essential proof that crazy transcends geographical and cultural boundaries. Indeed, this is where Fred Rogers got the phrase "Who are the crazy phuckers in your neighborhood?" though PBS ultimately made him change it to the less inflammatory (and substantially less accurate) "people."

Nobles

It's rare to get a moment with these folks, as they maintain several homes in their outer orbits—Windsor Island, St. Barts, Sagaponack—and their kids go to the uber-ritzy private day school two towns over. Even if your kiddo goes there too, chances are you won't be chatting up Mom or Dad Noble at drop-off or pick-up, as they're not inclined to bond with anyone. But if your kid is besties with one of theirs, he probably will score an invitation to a birthday party, the cost of which could easily pay off your own student loans and potentially your mortgage. While you're regaled with the details of how Young Master Noble 'coptered in to his party on the fifty-yard line at Sun Life Stadium, after which Flo Rida rapped him a "Happy Birthday," try to feel good about the birthday you just booked at Chuck E. Cheese, where you'll consider it a banner day if the rat high-fives your kid.

Kinetics

Unless your kids are on a team with theirs, the only time you'll see this family is when they're backing their minivan, packed with gear, coolers, folding chairs, and pop-up canopies, down their driveway en route to some athletic field. The members of this family have a highly inflated sense of competition, and each of the kids participates in a different sport, so they are rarely home. Instead they're huddled up at practices, private coaching sessions, or games/meets, where they bond readily with other team parents who share their views on coaching staff, tryouts, game strategy, college scholarships, and using tax dollars to build a new city stadium. Still, take care when mixing. Though most interactions

are neutral, a handful of elements in this family (known as "douche canoes") can react violently, especially when combined with three or four six-packs of beer on game days. This produces a "solution" known around the lab as court-ordered anger management.

Magnetics

Owners of "that house where all the kids play," this family attracts just about every element in the neighborhood with its "C'mon in!" open-door policy, finished basement, multiple game systems, and a pantry that resembles the snack-foods aisle at Target. When parents can't find their kids, they don't have to wonder where they are. They all gravitate here, where Mom happily pops another tray of frozen pizza nugget rolls into the oven, and Dad shows the kids how to do backflips off the roof into the pool.

Progressives

You'll know them by the rolled-up copies of *Mother Jones* in their totes, their pour-over coffees, and an ability to start any conversation with "You know, I heard on NPR . . ." As a family, they've marched on, sat in, stood up, and camped out, and they're naturally drawn to others with an overdeveloped sense of moral outrage. They will eagerly bond with anyone who shares the pet cause they're championing at the moment and can be talked into shooting out petitions from their emails like beams from a particle accelerator. These folks really start to sputter and fizz when they get rolling. You don't need to worry about making their acquaintance. They will find you, if only to capture your signature and email address. All the better to blast you with invitations to join them in—and donate to—their cause du jour.

Inerts

Found naturally at Phish and Jimmy Buffett concerts and splayed out on backyard chaise lounges, "really feelin' the Earth spin, man," the members of this family are so incredibly laid-back and "whatever, dude" as to be, well . . . inert. Dad's a master gardener, known for the "greenhouse effects" of the many varietals he's cultivating, while Mom's always tinkering with new recipes for cookies or brownies. Just don't let the kiddies eat them.

Au Naturels

Camping! Hiking! Sleeping under the stars! Cooking over a fire! Peeing on trees! Squatting in poison ivy! Getting dive-bombed by mosquitoes! This family is all about getting in touch with O_2 and H_2O and the rest of the Earth's elements. They never met a trail they didn't want to explore, a river they didn't want to raft, or a cave they didn't want to spelunk. No fancy RVs, log cabins or pop-up tents for these guys. They believe that eating off the land and sleeping on the ground at the mercy of wildlife builds character. Outdoor fun includes scat readings and identifying venomous insects and snakes. They're roughing it like survivalists. You know, in case the robots ever do rise up. When they invite your family to join them on vacation, make sure to book your own hotel room.

Law of Induction

The stronger the
arm-twisting
to recruit you into the PTO,
the more "volunteer" work
you'll end up shouldering.

Play Date Compatibility

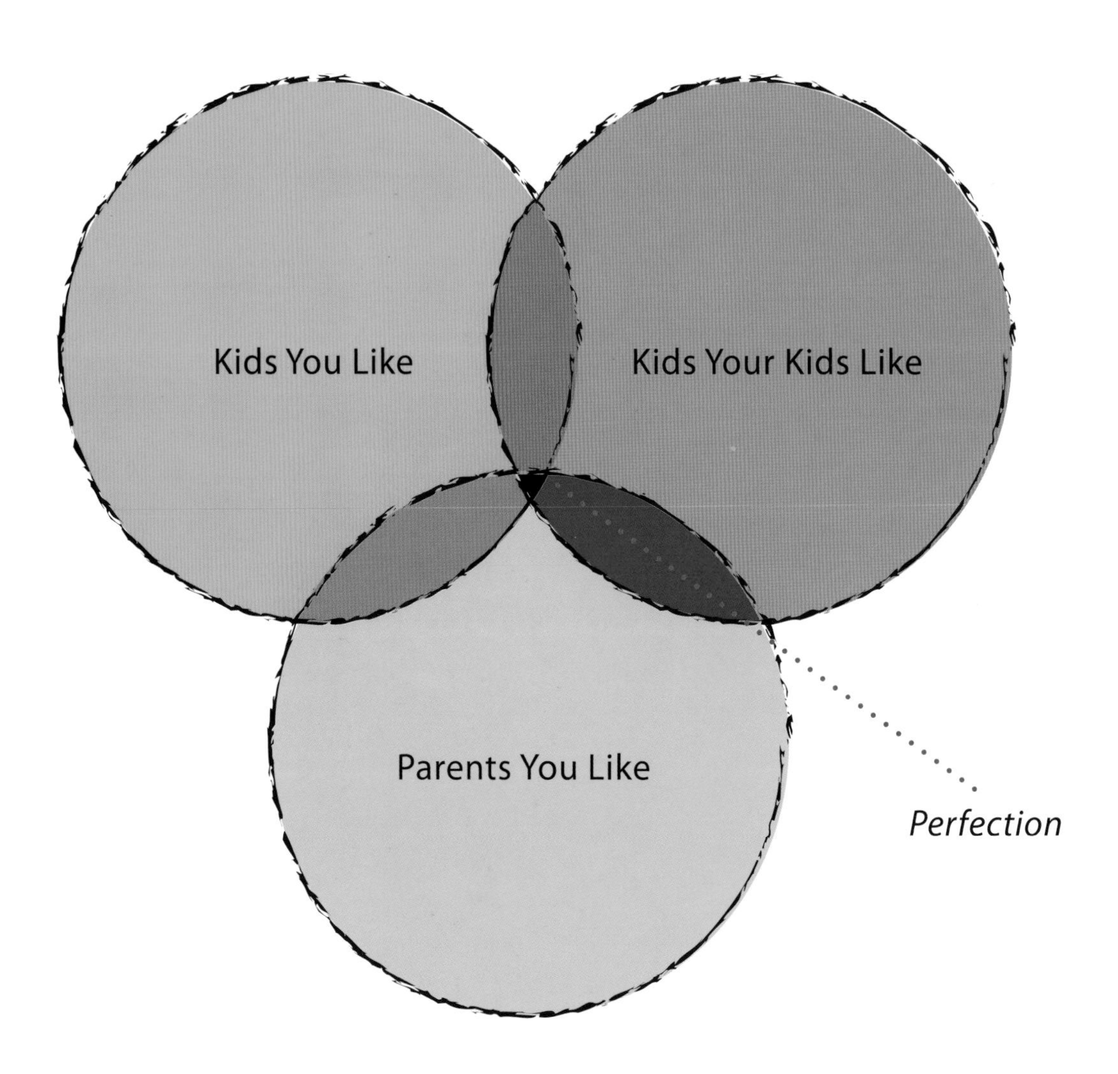

Lab Notes: Passive-Aggressive Parenting

Play dates and sleepovers are the ultimate lab experiment. Combine your kid with others, and you never quite know what might result, as each new friend is bonded covalently with a highly uncertain, potentially unstable element: his or her parent. Sometimes you get lucky—*They're normal!* And sometimes you end up with high-test crazy. Today's technology brings that insanity right up to your face—which is why we always advocate wearing safety goggles. There's really no element quite as volatile and reactive as a parent with a lightning-fast "send" finger.

Hey Amy!
Thank you sooooooo much for having Laney over on Friday night! She had SOOO much fun!! She was so tired she hid out in her room all day on Saturday! #momscore LOL!! Hey, did Livi find Laney's pink sweater by any chance? She's pretty sure she left it there. Next time, our house! - Kim

Hi Kim!!
Any time! They were up SUPER late, I heard them giggling at 2 a.m.! I hope that didn't totally screw up your day! Livi could NOT keep her eyes open today either, haha! I'll ask Livi to look for that sweater, I'm sure it's floating around somewhere!

Talk soon, Amy

2 FREAKIN' A.M. They were so damn loud! NONE of Livi's other friends are that loud. And who's idea was it to eat BROWNIES at midnight? Not Livi's. Never again.

Amy,
OMG not at all!!! I had peace and quiet all day! #totalheaven

Laney's pretty sure she left the sweater there, she can't find it anywhere. Just let me know when I can swing by and pick it up! Thanks again! - Kim

Thanks for NOTHING. Dealing with a cranky kid was JUST how I wanted to spend my Saturday! Never again. And I am pretty DAMN sure that sweater is stufffed under Livi's bed, right where she crammed it.

Kim,
Livi can't find the sweater anywhere :(Pink, right? Maybe it got stuffed in Laney's backpack?

Let me know, Amy

We don't have your cheap-ass pink sweater, OKAY? Why don't you take five seconds and look through Laney's stuff instead of jumping to accusations? That kid's a hot mess; she probably ate it.

Amy,
Nope, not in Laney's backpack :(Could you take a look?
It probably got shoved under the bed or dropped in
Livi's hamper.
IDK what these kids do with their stuff LOLOL! Cray-cray!
#INeedaMap #AndANap!! - K

Yeah right, what a surprise. This is the third item of clothing that has gotten "lost" at your house. Two words: LIVI STEALS.

Kim,
I'm sooooo sorry but I can't find it anywhere!! Livi doesn't even wear pink, so I would notice it for sure if it got mixed in with her
laundry. I'll keep an eye out though!

Amy

Holy crap, USE A WORD! LOL? IDK?? What are you, twelve? And WHAT is with all the hashtags? I've got one for you: #STFU

A,
Ok, well I'm sure it will turn up, eventually! Thanks for looking, and thanks again for having Laney over!
#You'reTheBest - K

#SayonaraSweater. Bitch.

K,
You bet! See you at Brownies on Thursday!

Amy

Bitch.

Hooke's Law of Parental Patience

Mom's last nerve

will streeeeeeeeetch

in proportion

to a child's annoying behavior.

Until it snaps.

mom...mom?
mom...mom...
mom...mom...mom...mom...
mom...mom...
mom...mom...
mom...mom?
mom...mom...
mom? mom...
mom...mom?
mom...mom...
mom? mom...
mom? mom...
mom...mom...
mom?
mom...mom?
mom...mom...
mom?
mom? mom

Chemistry Lab: Making Sparks Fly

Surveys show that after all the baby-making whoopee—and the long post-birth dry spell—dads want sex even more than court-side seats next to Jack or a Porsche GT3 with manual transmission. As it happens, moms want sex too (though the Porsche Panamera sedan is very tempting). But yes, moms do want sex. It's just that sometimes it takes a little sump'n extra to spark that reaction.

In chemistry, we call that extra little kick a *catalyst*. Catalysts change the activation energy thresholds required to facilitate the desired reactions—like fireworks or an explosion or a shower quickie while the kiddo's watching *Bubble Guppies*.

Consult the handy table we've provided for each catalyst's activation energy points. Plan accordingly.

Catalyst: Put clothes in laundry hamper.

Activation energy points scored: 5

Chance of sex: 10%

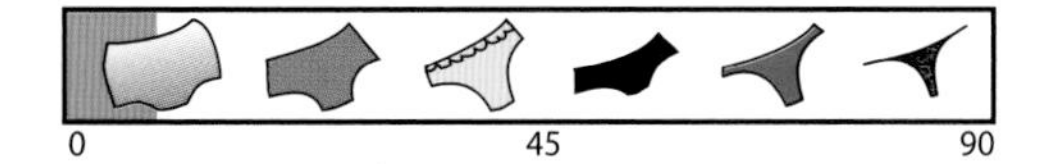

Catalyst: Do laundry.

Activation energy points scored: 20

Chance of sex: 20%

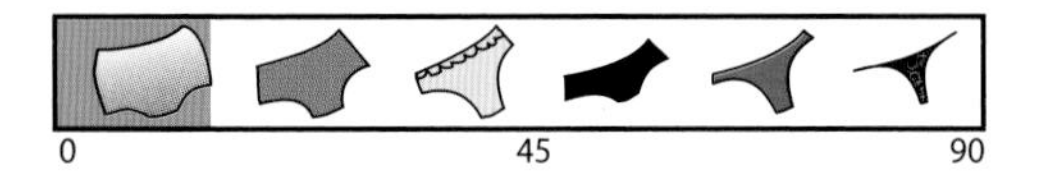

Catalyst: Do laundry, fold it, and put it away.

Activation energy points scored: 70

Chance of sex: 95%

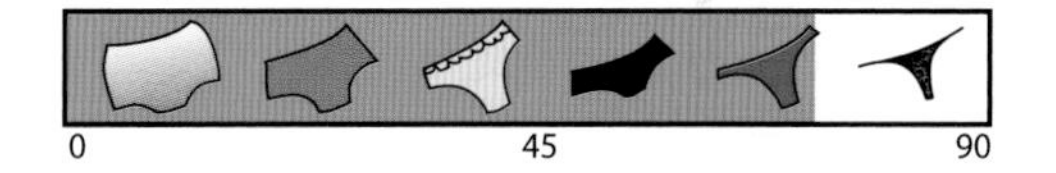

Catalyst: Greet Mom at the door with wine.

Activation energy points scored: 15 (per glass)

Chance of sex: 20% (with each glass consumed)

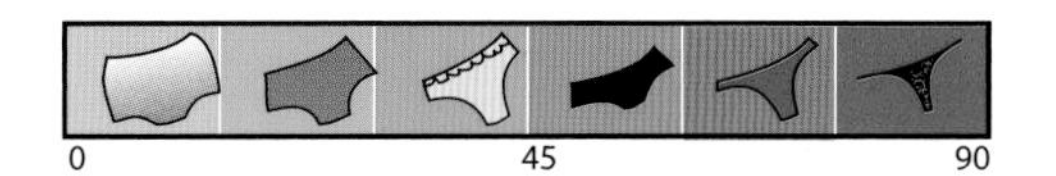

Catalyst: REALLY clean the house. Yes, bathrooms too. And floors. No complaining.

Activation energy points scored: 90

Chance of sex: 100%

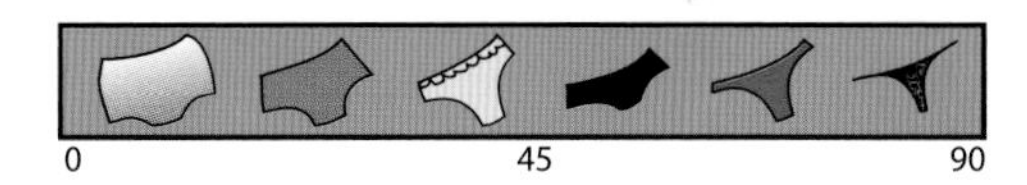

Catalyst: Start any DIY home improvement project.

Activation energy points scored: 40

Chance of sex: 45%

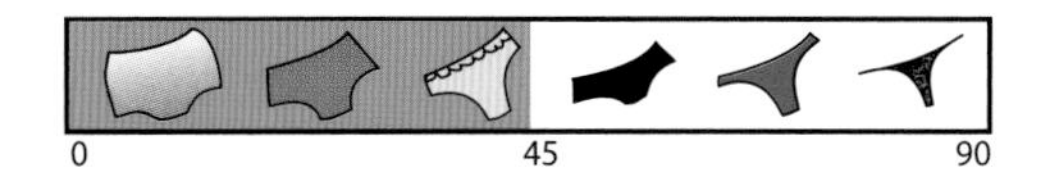

Catalyst: Finish it.

Activation energy points scored: 80

Chance of sex: 90%

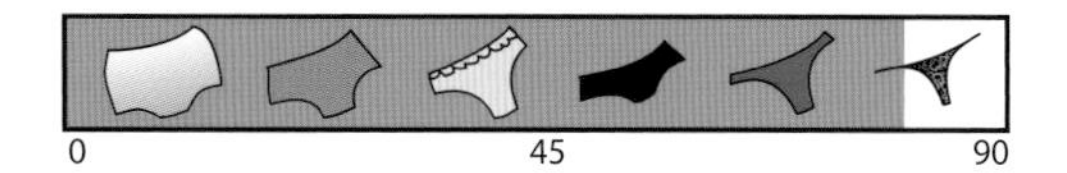

Catalyst: Go grocery shopping.

Activation energy points scored: 50

Chance of sex: 55%

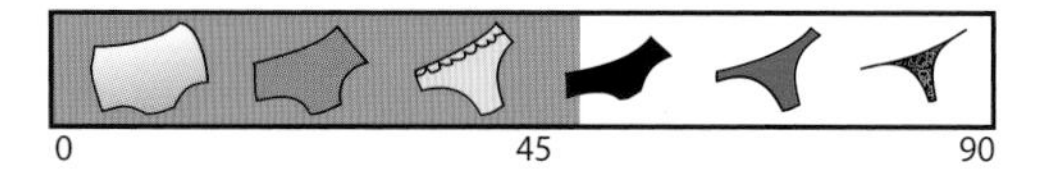

Catalyst: Come home with something besides Funions and beef jerky.

Activation energy points scored: 70

Chance of sex: 85%

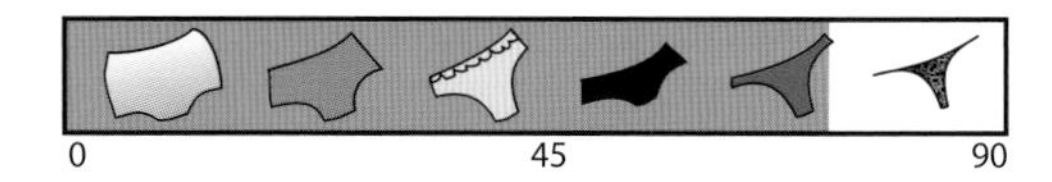

Catalyst: Take Mom out to a restaurant without a kids' menu or a playscape. Or kids.

Activation energy points scored: 50

Chance of sex: 70%

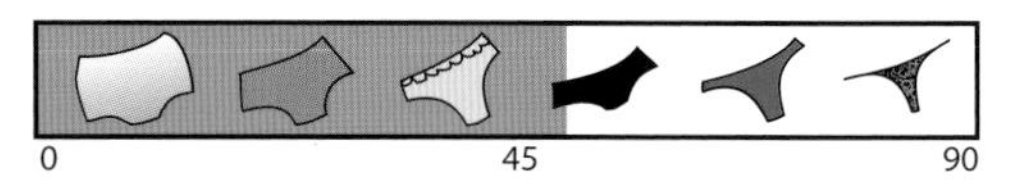

Catalyst: Take Mom away for the weekend without the kids.

Activation energy points scored: 80

Chance of sex: 90%

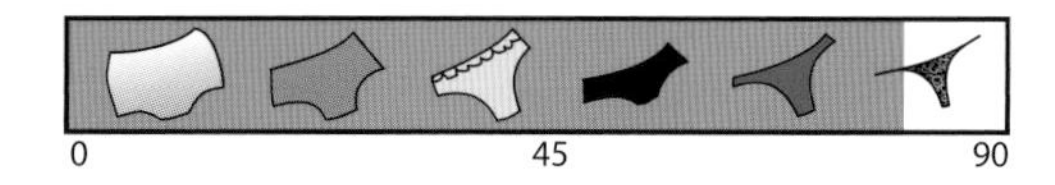

Catalyst: Send Mom on a weekend away by herself.

Activation energy points scored: 90

Chance of sex: 100% (if she comes back)

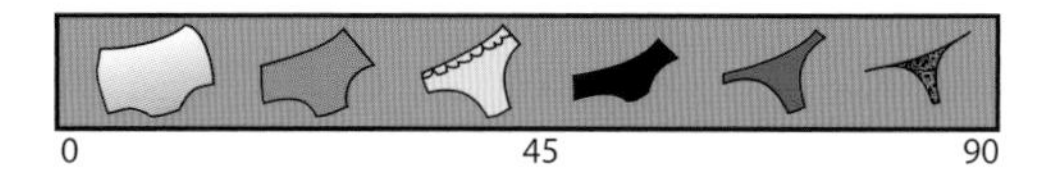

Beverage-to-TV Index

Recommended minimum beverage consumption to get through your child's favorite TV programming with your sanity intact.*

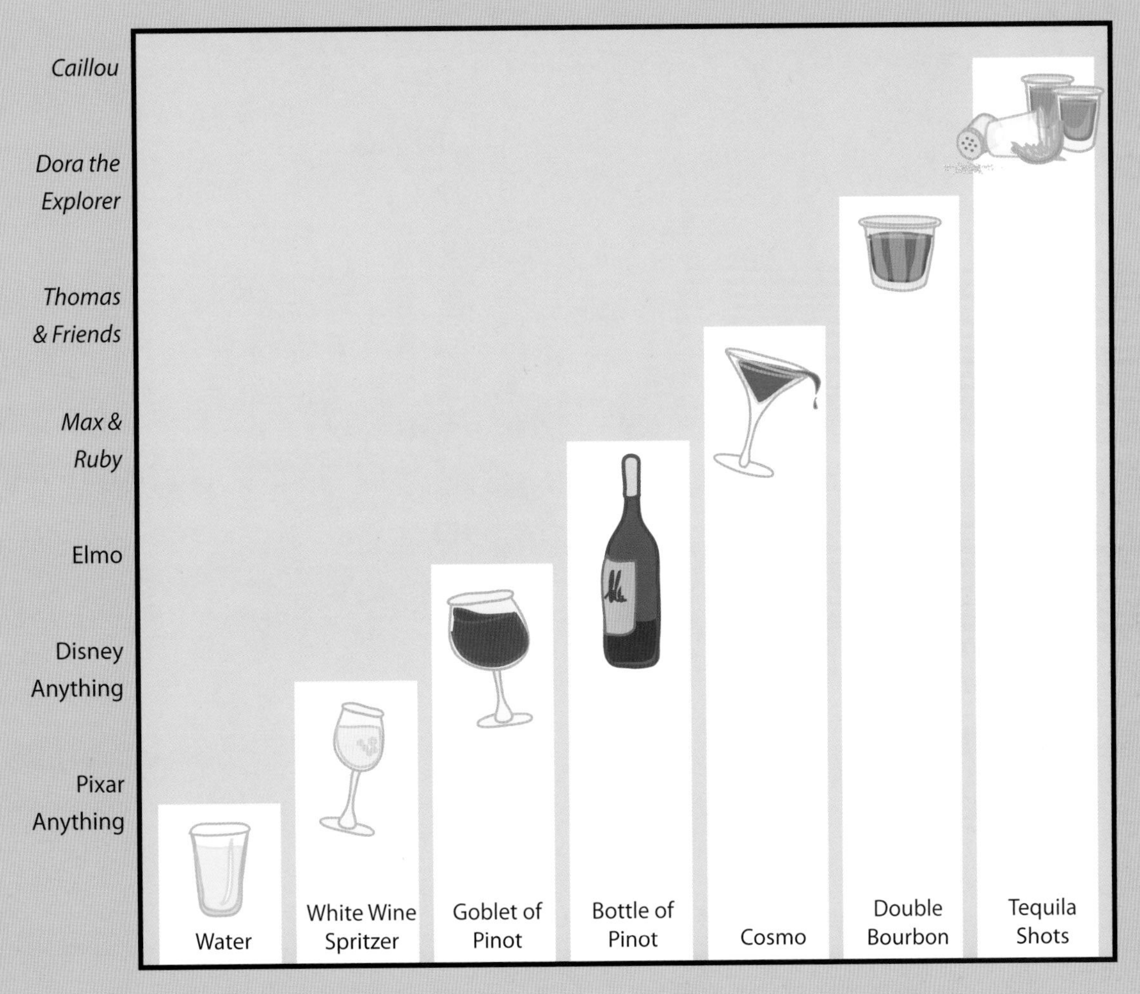

* Your mileage may vary.

THERMAL ENGINEERING

Parents perform impressive feats of science every single day, taking raw materials—organic matter, ionic compounds, oleic acids—adding heat, and *voila!* A patty melt!! It's like you're Marie effing Curie. (Though hopefully not so radioactive.)

But what of the culinarily *un*-inclined, those of us for whom cooking doesn't extend much beyond, "Pierce plastic wrap with fork. Microwave four minutes on high"? Not to worry. Seriously. Your kitchen could get three Michelin stars, and your kid still wouldn't eat what you worked hours to prepare.

For reasons that continue to perplex scientists, babies will eat an exotic variety of pureed foods, but sometime after kids discover the word *No*, their palates shrink like 100 percent cotton T-shirts in the dryer. From then on, kids exist on mac and cheese, French fries, chicken nuggets, hot dogs, dry Cheerios, PBJs and the occasional grilled cheese. But only if the sandwich is made with cheese that is both orange *and* square. No exceptions. Trust us on this. Competitive lunch boxing is a sport for moms. Kids? They're (mostly) happy as long as you keep the noodles with fluorescent orange cheese coming.

Food-Loss Index Calculation

The method for determining
the total amount of food
you must prepare
based on what will
actually end up in your toddler's mouth
minus what ends up on his clothes
. . . his hair . . . the floor . . .
the walls . . . you.

General Child-Feeding Protocols

Once kiddos start solids, meal times can become wonderful lessons in culinary exploration, as parents introduce them to the deliciously wide world of foods beyond breast milk and formula: Oatmeal! Pureed prunes! Bouillabaisse!

Still, however enthusiastic parents are to acquaint children with baby octopus and chicken tikka masala (so that they can finally choose a restaurant based on parameters other than *Does it have a kids' menu?*), they must still be mindful of essential food-safety guidelines. Because of the very nature of enticing a child to try new foods, meal times, while filled with endless adventure and teachable moments, can also be filled with frustration, rage, and the kinds of stains that won't come out of clothing or curtains no matter how many times the dry cleaner tries.

The following protocols can help minimize the hazards associated with feeding children. Learn and follow them to ensure a safe dining environment for you, your children, and those around you trying to eat without getting caught in the crossfire.

1. Food must not be "weird" in any way. "Weirdness" to be determined at the child's sole discretion.
2. Favorite foods are subject to change without notice. Prior consumption is no guarantee that the same food will be eaten again.
3. Requests for a specific food do not guarantee that food will actually be consumed.
4. There should be no expectation that a child will eat what everyone else is eating. Individual meals will be prepared on demand.
5. New foods will be regarded with extreme prejudice. Requests to serve anything new must be submitted in advance, in triplicate.
6. One food may never touch any other food on the plate AT ANY TIME.
7. Food will only be eaten if served on the "right" plate. Parent has no say in "rightness."

8. The ratio of French fries to any green vegetables on a plate must always be 8:1.
9. Minimum number of bites eaten must be negotiated in advance. No exceptions. (Except for cake.) Size of bites taken to be determined at the child's sole discretion.
10. Triangles are the only acceptable shape for sandwiches. Sandwiches cut into squares will be deemed "ruined" and must be made again.
11. Fruit and vegetables must be presented in adorable animal tableaux as can only be found on Pinterest. Child's delight in seeing animals painstakingly constructed of bananas, watermelon, and oranges does not constitute a binding

agreement to eat said fruit creation, no matter how delicious the parent claims it to be.

12. Ketchup must be used generously. Ditto ranch dressing.
13. Unwanted food may be placed on a parent's plate, even after it's been chewed.
14. Forks will not be used when fingers will suffice.
15. Napkins will not be used when shirt sleeves, shirt fronts, or pants legs are available.
16. Being full after three bites of dinner does not preclude requests for a PBJ at bedtime because the child is "starving."
17. Prepackaged snack foods must be opened, but NOT ALL THE WAY. Completely opened snacks will be rejected, and a new snack must be presented.
18. Denial of dessert will not be tolerated. Any attempt to refuse dessert will be met with severe tantrums.
19. Broken cookies are "tainted" and will be summarily refused.
20. Boogers *are* a food group.

Dinner Prep-Time Estimation

The time it takes to toss a salad together or throw some spaghetti in a pot is dependent on the number and size of the hands that insist on "helping."

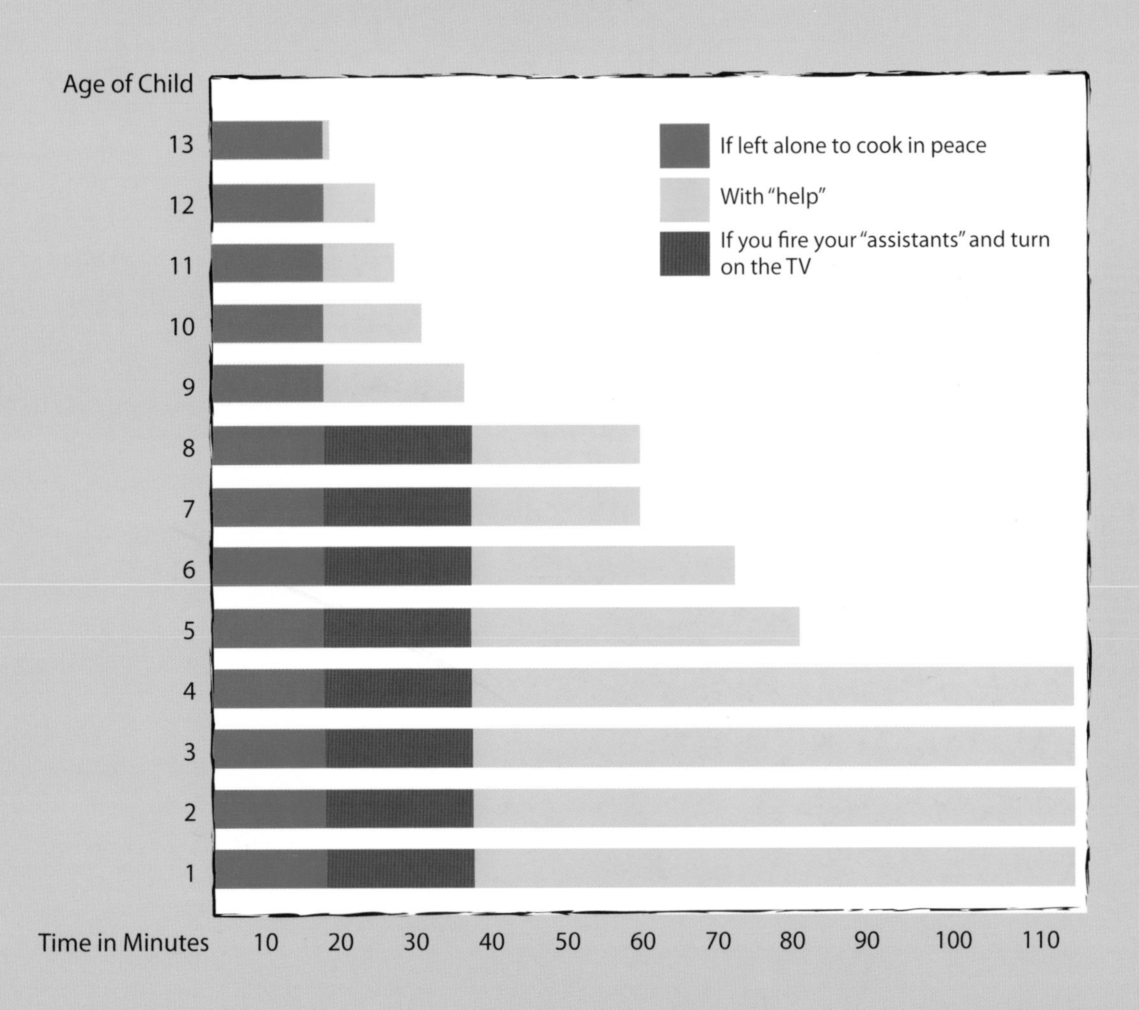

The Periodic Table

The place family meals
are served
and where your child will
occasionally stop by
to nibble from his plate
throughout the evening.

Experimental Gastronomy: A Study in Potatoes

Objectives: To determine if a child who eats French fries by the pound will enthusiastically consume potatoes when they are served in other forms, specifically mashed.

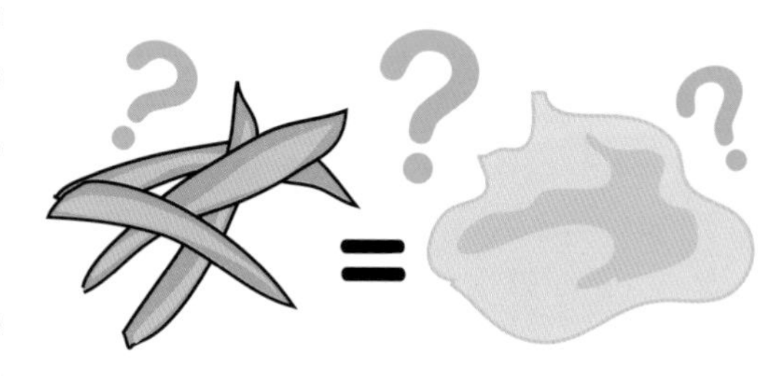

Population: One extraordinarily finicky eater, age four. One non-picky eater, age four, to function as the control group.*

Methods: This trial was conducted at 5 p.m. (aka dinner time) in the kitchen at the preschooler's home. Prior to the experiment, there was a three-day run-in period, during which the preschooler was served French fries with ketchup alongside her main course to verify that French fries were acceptable and palatable to her. No other potatoes were served at dinner time, and the preschooler did not consume potatoes during any other part of the day.

---

* The research team was unable to find an age-matched control who was not a picky eater, so the four-year-old family golden retriever "volunteered." (And by "volunteered," we mean "begged at the table" till we fed him whatever the preschooler refused to eat.)

Data Collection: During the experiment, fries were replaced with a generous scoop of fluffy mashed potatoes (liberally seasoned with salt and melted butter) and presented to the preschooler. Preschooler seemed surprised by the new food. And surprise quickly turned to alarm, then rage. Preschooler was observed shoving the plate away, pursing her lips together, and flashing the researcher the universal sign for Talk to the Hand, Byotch.

Researcher made a second attempt to engage preschooler in the mashed potatoes, pointing out that preschooler loves French fries and that, in fact, mashed potatoes are exactly like squashed fries. Preschooler steadfastly refused to sample the mashed potatoes.

On a third attempt, Researcher made airplane sounds and said, "Open up the hangar! Here comes the airplane! Yummy!" Preschooler's opposition to the mashed potatoes remained firm.

On a fourth attempt, researcher tried the Just Take One Little Bite tactic, offering assurances that the mashed potatoes need not be finished, only "tasted" or "sampled." This was met with a firm "I DON'T LIKE MASHED POTATOES!" Researcher countered with, "How do you know you don't like it if you haven't tried it?" Preschooler proved immune to logic.

On the fifth and final attempt to persuade the preschooler to taste mashed potatoes, the researcher changed tactics completely, offering a bite-for-bite exchange, mashed potatoes for Hershey's chocolate kisses. Preschooler appeared to consider the exchange

(giving rise to the hope that the mashed potatoes *might* be consumed), but then refused.

Mashed potatoes were fed to the control group, which licked the plate clean.

Results: During an experiment that tested the willingness of a four-year-old finicky eater with an affinity for French fries to try a potato preparation variation, preschooler refused all attempts to persuade her to try the mashed potatoes on the grounds that they were not French fries. Control group demonstrated no reservations about eating mashed potatoes (or any food) presented on a plate. However, researcher wonders about the control group's standards for palatability, as control group is also known to lick its own butt.

Conclusions: On the basis of this study, researcher concludes that preschooler is unlikely to ever eat potatoes when not presented as French fries. Given the ubiquity of picky eating among children under thirteen, researcher believes these results will apply to other populations. Parents should be aware that when potatoes are served in any form other than fries, children are likely to dispose of their dinners by feeding them to the ~~dog~~ control group.

Where Your Kid's Food Goes

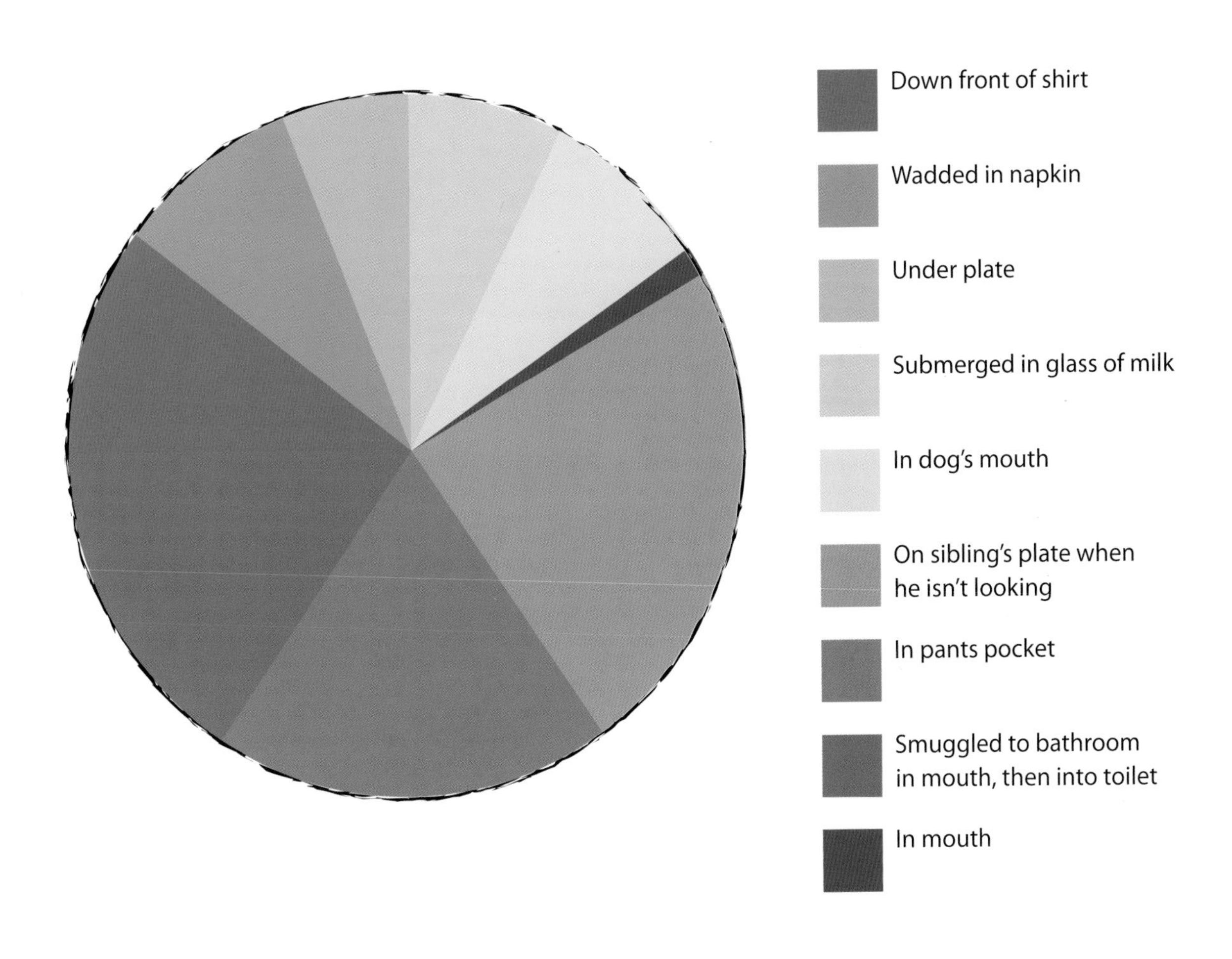

Brownian Motion

The burst of kinetic energy
produced when a child
gobbles down a double-fudge
brownie
right before bedtime.

The Four-Color Problem

Persuading your preschooler
to settle for a chocolate sheet cake
for her birthday . . .
after she's spied a
four-layer ombre cake.

COLLISION FREQUENCY

Chemistry teaches us that the more molecules we have in a tightly confined space, the more collisions will occur and the more reactions will result.

If that doesn't sound like a bounce house jammed with sugared-up tots at a birthday party, we don't know what does. Somebody cue the H_20 . . . uh tears! (*Bah-dum tish* . . . We'll be here throughout the book . . . Don't forget to tip your cocktail servers! Take my kid! Please!)

All kidding aside, kids' activities provide abundant opportunities to observe the phenomenon of "collision frequency" up close: soccer scrums, basketball games, flag football, softball, tag, sword fights, pillow fights, Nerf blaster fights, sibling squabbles, aggressive My Little Pony play. Even the meet-up of toddler cheek with table edge. And . . . oooh, Ow! Yeah. That last one's gonna leave a mark.

All the better to acquaint you, m'dear, with your friendly neighborhood urgent care center. Now, can somebody take my kid, please? I think he needs stitches.

Critical Mass

The maximum number of kids
a parent can supervise,
solo,
before his head explodes.

Play Group Chemistry: Five Elements Your Kid Will Surely Mix With

Take a look around you. G'head. Take a really good look. Okay, so everything that you see, the furniture, the sleeve of Chips Ahoy! you're eating before the kids get home, the cat that just vomited all over your freshly shampooed carpet, even the dust bunnies tumbling under the kitchen table that you hope *no one* sees, all of that and everything else is made up of some combination of elements in the periodic table. Pretty cool, huh? We think so too. Now, some elements, like gold, silver, and copper, are as natural as dirt and have been around just as long; others, like Livermorium and Ununoctium, were cooked up in labs within the last few years. All told, there are some 118 elements on the periodic table, and the race is on, like some quantum chemical scavenger hunt, to discover still more.

Still, we were sort of surprised to hear that American scientists believe they've already discovered *five* additional elements. According to the researchers, these elements naturally occur around playgrounds, daycares, scout troops, peewee sports, dance classes, and in exponentially high numbers around Disney World and Sesame Place.

Of course, two labs need to confirm an element's existence before it can be named by the International Union of Pure and Applied Chemistry (IUPAC) and added to the periodic table. If added, the following five would form the element family, the Play Group.

Bullshitium

Much like "fool's gold," this element appears just a little too incredible to be believed. It's often "said" to be found on remote private islands, Hollywood sound stages, music videos, the third baseline at Fenway, and wherever Air Force One is docked at the moment. But in actuality, it can be scooped up from the same park sandboxes as most other elements. It's quite benign, but safety goggles are recommended around Bullshitium, as severe eye-rolling injuries can occur.

Gamerium

This element is so stable, it's practically inert. Scientists located it hunkered down in front of PlayStations or Xboxes or Wii's—but not the games that require any actual, you know, physical exertion. Occasionally, Gamerium can be coaxed to interact with others, though typically it requires a digital matrix, such as Minecraft, Plants vs. Zombies, or Clash of Clans, to do so.

Psychotium

This element is unstable, highly toxic, and, depending on the other elements present, can be fairly unpredictable when added to the play-date mix. Psychotium has the *potential* to blend without incident in the right combinations. However, it is known to be combustible and requires very little to set it off—e.g., sandwiches with crusts; drawing the gumdrop card in Candy Land. It's worth noting that terror groups are always looking to acquire more Psychotium. Handle with extreme care.

Dangerium

When it comes to wild, crazy experiments, this one is the preferred material for pushing the limits of velocity, force, and matter. Which is exactly why it's most often found at the tops of stairs and the tallest slide at the playground, in the highest tree branches, at skate parks, on snowboarding trails and in major medical centers' emergency rooms. Fortunately, it reacts very well with fiberglass and Dermabond, which makes it fairly easy to repair.

Dramatium

Safety goggles are a must when handling Dramatium, as this element is known for the colorful showy sparks it emits when agitated. Though not toxic, the tiniest of disturbances—green grapes; the "wrong" shirt; getting a juice box with the straw already in it—can cause sudden explosions. Fortunately, Dramatium's flames burn hot and fast, typically fizzling out once it runs out of energy. Though it may sulk the rest of the day.

Parental Rates of Reaction

The intensity of a parent's reaction to any given child's antics is strongly correlated with the age of the child in question.

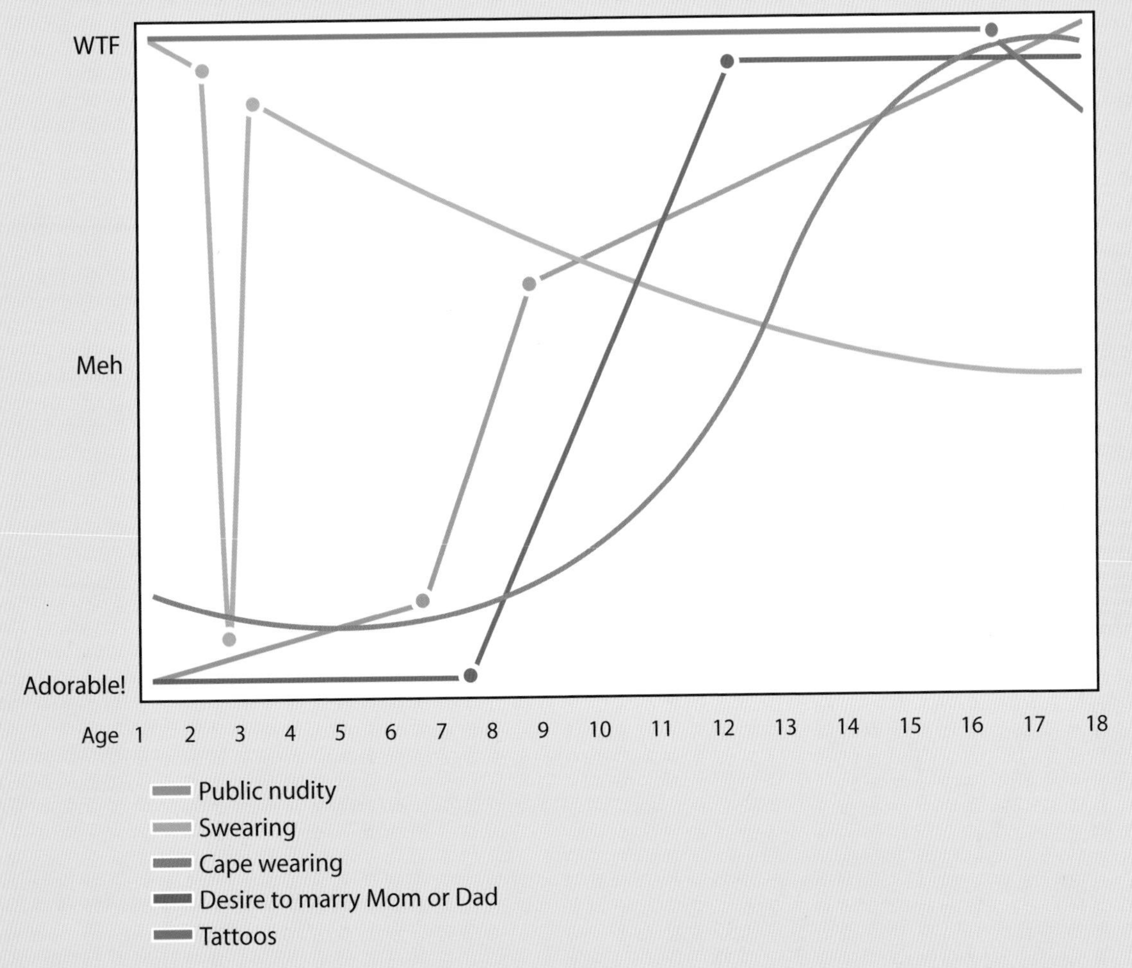

Fraternal Combustion

The temptation to allow siblings
to punch it out and finally settle it
"once and for all."

The Sporting Life: What's in It for Us?

Bazoodles* of research validate the benefits of sports for kids: Kids who play sports are stronger physically and academically and less likely to end up running cocaine for a Mexican drug cartel. Which is why we are up at the ass-crack of dawn on weekends, shuttling them to practices and games two zip codes over. Still, it would be nice if parents got something for our efforts, apart from sun damage from baking on the sidelines, cheering them on. So, here are a few sports and activities we'd like to see added to the roster.

Track and Parking Lot

Drop-in weekend clinics for all ages. Coaches are stationed at your supermarket parking lot from 10 a.m. to 4 p.m. for hour-long practices focused on cross-lot wind sprints, cart hurdles, and increasing players' agility through an ever-changing vehicular obstacle course in order to help Mom locate the car when she's forgotten where she parked.

---

*Increment of measure larger than a googol.

Capture the Can

In this variation on Capture the Flag, participants fill the garbage cans and recycling bins with the appropriate refuse, haul them out to the curb on pickup days, and then retrieve them once they're emptied. Whoever gets the right household waste into the correct container and out to the curb on the correct day wins. Events held rain or shine.

Competitive Weeding

At this neighborhood-wide sporting event, participants are divided into teams and set loose to collect the most weeds from all flagged yards. The team with the largest collection of weeds will be declared the winner, thereby earning a five-minute head start at the next week's event.

Search-and-Recovery Marathon

In this scavenger hunt meets long-distance run, each kid is given a list of all of the shoes, socks, jackets, backpacks, books, electronic devices, sweaters, and thermoses that have been lost over the last quarter and then races to retrieve them. The winner is the one who returns with every last item on the list. Losers get the value of the lost items deducted from their allowance.

Fold n' Stack

This sport involves folding laundry, then shifting and re-shifting the piles to achieve the tallest, most stable stack that can successfully be carried upstairs without incident. Each stack must be grouped by clothing owner. Bonus points given to players who put their clothes away rather than just dumping them on the floor.

Scooper Yoga

This style of yoga helps develop flexibility and core strength as kids learn to bend and stretch while scooping dog poop in the yard or cleaning the litter box. Children who adore animals and are a bit OCD tend to excel at this yoga style.

Lights-Out Dash

A series of fun runs, this event requires that participants dash through the house, turning off every light that's been turned on before leaving the house for school or to run errands. Winners are announced at the end of the month when the electric bill arrives.

Kitchen Curling

A close cousin to Olympic curling, this is a team sport in which each child on the team is given a broom and a floor to sweep. The one with the biggest pile of fur balls, dust, crushed Cheez-Its, and mini erasers in his or her dustpan wins.

Target Relay

Parents should come prepared each week with a Target shopping list of at least ten . . . oh, who are we kidding? . . . make that twenty items. Teams of two will relay-race from the in-store Starbucks out into the aisles to retrieve each item, while parents sip lattes and goof around on their phones. Bonus points will be awarded to the team that successfully uses coupons and checks out without adult assistance.

Contributing Factors

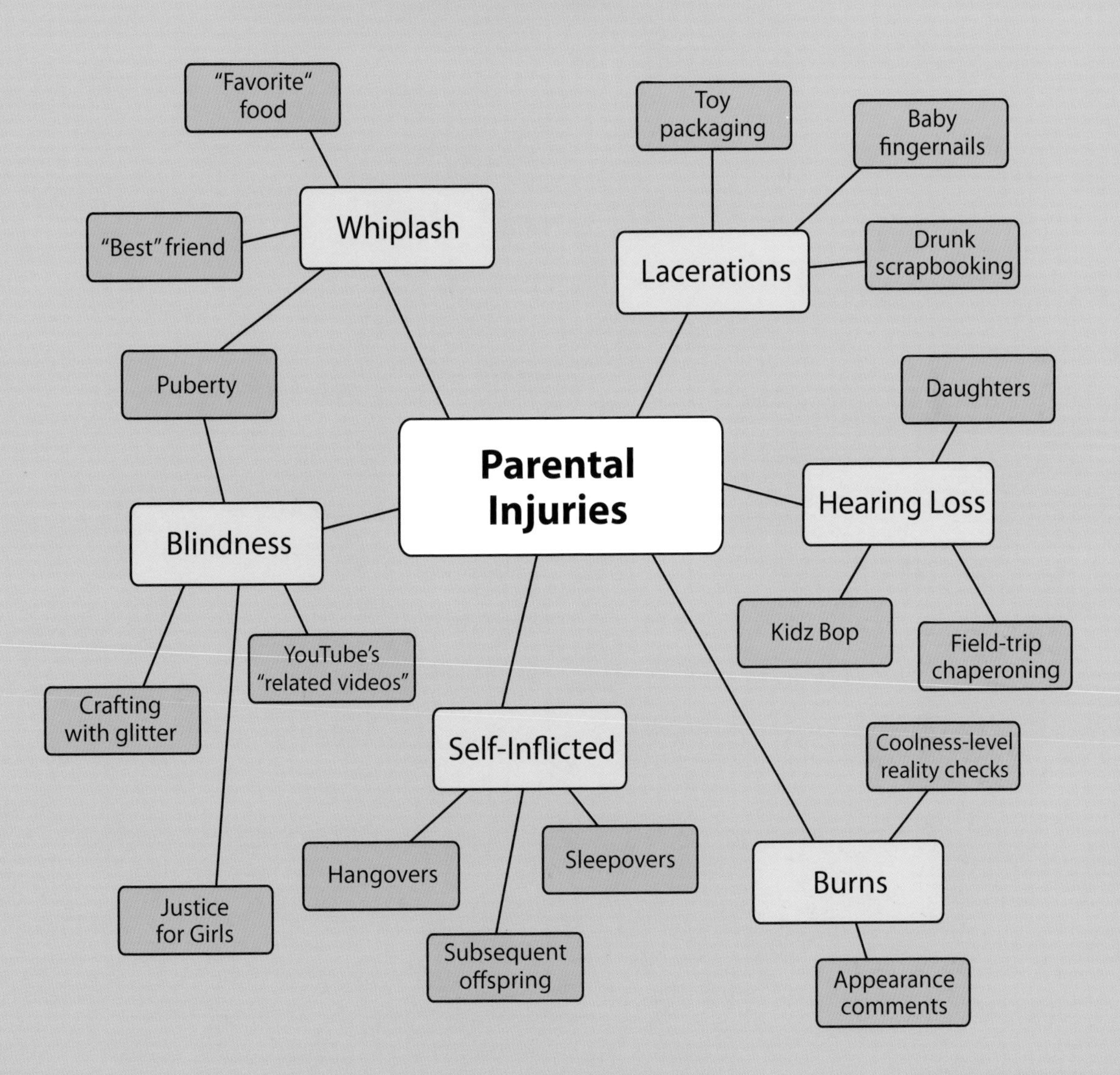

Ideal Gas Law

The child who lets loose a
particularly resonant burp or
a resounding fart
MUST announce it to everyone present
and then duplicate it . . .
repeatedly.

BURP!
BURP!
BURP
BUR
BURP

"Life would be tragic if it weren't funny."

- Stephen Hawking
Physicist

Physics

ENTROPY

IS JUST ANOTHER WORD FOR

I CAN'T FIND A THING IN THIS MESS

"Things fall apart; the centre cannot hold; mere anarchy is loosed upon the world . . . " Newly discovered letters between the poet William Butler Yeats and his friend and fellow poet Ezra Pound reveal that Yeats scribbled down that line in utter frustration on a day when he was trying to compose the poetry that would eventually net him a Nobel Prize while simultaneously watching his kids. The beleaguered Yeats wrote:

> *Dude, I'm going crazy over here. The house is a mess. I lost my favorite pen. Every time I get a line straight in my head, one of them interrupts me for milk or a piece of fruit or to tell me about some goddamn butterfly on a leaf, and poof! It's gone. Gone, man. I can't take it. It's anarchy over here. Can I work at your place tomorrow?*

We feel you, buddy. Chaos, mayhem, pandemonium. Call it what you will. As any physicist will tell you, the universe tends toward disorder, and unfortunately, that precept holds whether you're looking at a distant galaxy, reflected back from the Hubble Space Telescope; your home, buried under layers of toys; even your own mind where . . . um . . . wait . . . what were we saying?

Clutter, both mental and physical, comes standard issue with kids. Try though we may to contain it, entropy will out. Fortunately, research suggests that most parents do manage to dig out once their offspring depart for college. Until they return with *their* kids.

Destabilizing String Theory

Why those "fun" craft projects
usually end in a
tangled mess and tears.

Law of Urinary Dynamics

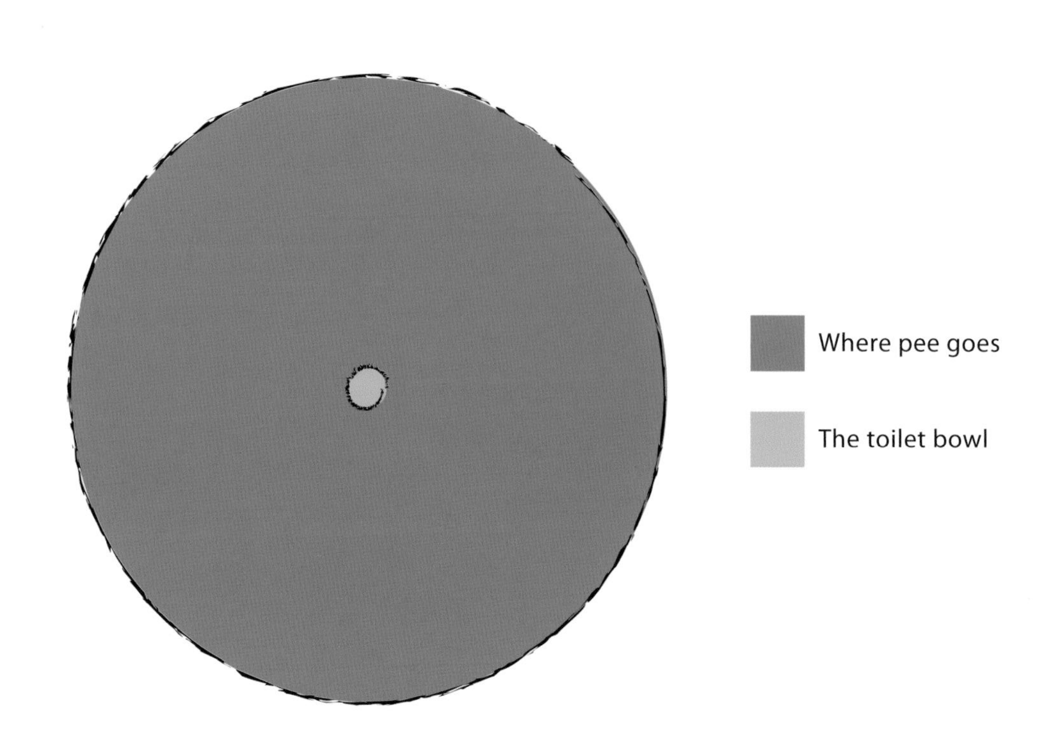

Approximately Three Minutes Inside a Busy Mom's Head*

The American mom brain is a frenetic thing to behold. Even when she's sitting still, seemingly idle, the hyper-caffeinated mom mind is always pop-pop-pop-pop-popping, like thousands of buttery popcorn kernels microwaved on high. *What needs signing? What needs baking? What needs finding? What needs fixing? What needs buying? What needs checking? What needs washing-drying-folding?* It's semi-organized chaos.

In an effort to quantify just how active the mom brain really is, we partnered with Johns Hopkins neuroscientists on a study in which we monitored a mom's thought process as she waited to pick up two kids from school. We chose car-line pick-up for our study because anecdotal research suggests that those three to five minutes, spent waiting for exhausted teachers to load cranky kids into minivans, is the only time of day when a mom is not engaged in any critical parental activity, thus allowing her thoughts to roam free.

---

* Hat tip to comedian Jason Good whose blog post "Approximately Three Minutes Inside the Head of My Two-Year-Old" never fails to make us laugh.

Here's a snippet from the raw data file:

Whew! Made it! They're not even loading kids into cars yet. HA! Someone behind me! I'M NOT LAST! Slacker!

Who's waving at me? Maya's dad! (Waving back) *Mmmm . . . hottest dad at school . . . How do I look?* (Quick surreptitious glance in mirror) *Is that a cheese puff in my hair? Lovely.* (Scribbles on hand: No more cheese puffs!)

(Absently runs hand up leg) *I should shave my legs. Hmm, what month is it? February? Eh, I've got time.*

CRAP! Library books! WHERE are they? (Rummaging in back seat) *Okay, here they—OHH-EWWW! WHAT'S THAT?!? WHAT MOVED? Oh, just a wad of hair from Zoe's brush. Ick. WHOA! What's that? Roaches?! Oy. Chocolate almonds all over the floor.* (Scribbles on other hand: Clean out car. TODAY!)

Did I sign that field trip permission slip? Which kid has a science fair project due again? (Scribbles on gum wrapper: Buy poster board.)

Do the kids have clean uniforms for school tomorrow? Probably in the hamper. Maybe I can Febreze them.

What's that say? Spirit night? At a vegan raw foods joint? WTH, PTO? No way I'm eating that. (Scribbles on crumpled napkin: Pick up tacos for dinner.)

(Turning up radio) "I bless the rains down in Aaaa-frickah! Gonna take some time to do the things we never had . . . " *Jeez, that song's old. Came out in high school. Gawd, now I feel old.* (Another look in mirror, scribbles on crumpled napkin: Find better anti-aging cream. And concealer. And schedule Botox.)

Finally, line's moving . . . C'mon people! Let's move! I still have work to finish! I wonder if I can get an extension . . . JEEZ! How long does it take to buckle a kid into a car? Gawd, I'm tired. And hungry.

Just what I need, Susan with her effing PTO clipboard. Effing sanctimonious bit—"Hey Susan!" (Rolls down window) "I'm good. How's your daughter? Four dozen? Chocolate chip? Sure, Susan! Count on me." *DAMN! Now I'm stuck baking cookies. Still, cookies do sound good. I bet the Girl Scouts are still at the grocery store. THIN MINTS!*

Wait. Why are all these kids wearing pajamas? Pajama Day was today?! I really should look at the school calendar. (Scribbles on car visor: Check school calendar.)

Just a few more cars . . . Where's my ID placard? Where'd it go? Crap. I must've tossed it. (Scribbles in dust on dashboard: Get new placard.) *I need a vacation. Or Thin Mints.*

Thin MINTS! Thin MINTS! Thin MINTS! Thin MINTS! THIN MINTS! THIN MIN—

(Door slam) "MOM! Where's the iPad?"

Thin MINTS! Wait. Where am I going?

Schrödinger's Backpack

The moment in which your child's
homework
simultaneously exists
in both a completed state
and a
"Holy-shit-it's-10-p.m.-you-haven't-
even-started-
and-it's-due-tomorrow!"
state.

Where Is Your Phone?

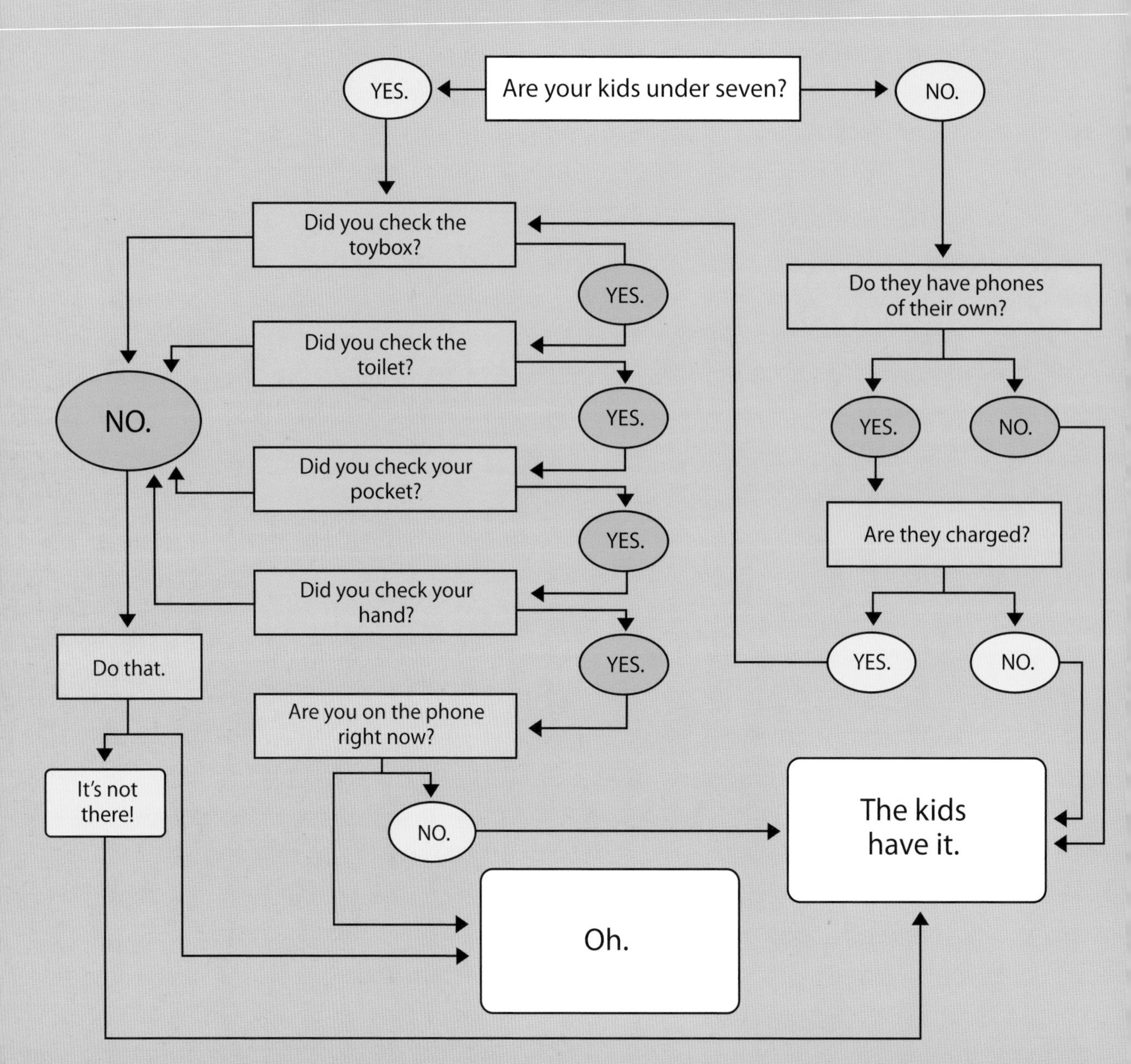

Your Home: The Perfect Proof of the Second Law of Thermodynamics

German physicist Rudolf Clausius was no doubt witnessing the swift, steady destruction of his home at the hands of his six children when he formulated the second law of thermodynamics, which states that in a closed system (a house with kids in it, for instance), disorder and chaos will naturally reign supreme.

Initially—that is to say, before Kid Number One really starts to get mobile or acquire any siblings—parents may delude themselves that it's possible to stem the red-blue-yellow plastic-toy tide or at least do a clean sweep at the end of the day.

But before long, as the blocks, crayons, Barbies, sticker books, bouncy seats, jumperoos, ride-alongs, play kitchens, board games, and wooden puzzles accumulate, the house will look like a typhoon tore through it and its name was Fisher-Price. By this time, even the most stalwart neatniks with pathological OCD have to admit: Clean-up is futile.

Stage One: Delusion

In the early months, B.C. (Before Crawling), Baby stays where he's put, and his world is largely confined to a generous square of baby blanket and perhaps a small bucket or basket containing the squishy, crinkly, fuzzy toys that entertain him. As this stage can last for the better part of a year, parents tend to get lulled into a false sense of *What the hell are people complaining about? Whiners. This isn't messy. I got this.*

Stage Two: Shock

Reality kicks in about the time a toddler works the top off a family-size jar of Vaseline, grabs two greasy fistfuls, and smears every surface within reach. Meanwhile, anything not covered with greasy goo has been spared only because it's buried under several layers of primary-colored plastic. This stage is especially hard on the home-decoristas who find that kid crap clashes radically with their I'm-Finally-A-Grownup decor of cool grays, taupes, and blues.

Stage Three: Defiance

Parents often attempt to reclaim their house, or at least declutter the living room. Swooping in, like feng shui ninjas, moms try to corral the toys in a toy chest or playroom, separating "outgrown" toys to be donated to Goodwill, while dads double-down on child-proof cabinet locks and place the Vaseline on a shelf far, far out of reach. But ohhh, if you've never attempted to purge the toys a child doesn't play with, be warned. Purging "unwanted" toys is considered the most serious breach of parent trust (matched only by serving broken cookies at snack time). Remember: Nothing sparks a mega meltdown over a toy that's never been played with like the threat that another child will get to KEEP it. There's only one acceptable course of action: Return all the toys to their original places.

Stage Four: Expansion

Finishing the basement, then offering it to the children as a dedicated KidZone can be a last-ditch effort to contain the game systems, jumbo snack packages, model airplanes, terrariums, and gaming chairs. But alas, several hundred extra square feet only spreads the mess further.

Stage Five: Retreat

The breaking point for parents comes when they finally realize there is no digging out, no remediating the home. The toys may change, but the clutter remains the same. Stuffing only what they can carry in the kids' Darth Vader and Hello Kitty backpacks and the twenty-three reusable grocery bags stashed in the pantry, parents list the house as a superfund site, then hightail it outta there as fast as their colorful Chuck Taylors will let them.

TRAVERSING THE SPACE-TIME CONTINUUM

If there's one scientist that parents really, truly get, like on a completely visceral level, it's Albert Einstein. There's a man who understood that time can do some wacky things—especially when you're a parent.

Take how he explained his theory of relativity to the public: "Put your hand on a hot stove for a minute, and it seems like an hour. Sit with a pretty girl for an hour, and it seems like a minute. That's relativity."

Seriously, what parent hasn't made a similar observation? One minute you've got a newborn, straight out of the chute, covered in goo and wrapped in one of those striped hospital blankets. Then you blink and you're watching a kindergartener skip into school without so much as a kiss goodbye or a backward glance. Then time moves so fast, you could cry. But, trapped at home during yet another snow day in an endless winter or in a sketchy gas station restroom waiting for a three-year-old to poop, time s-l-o-w-s to a glacial pace. That'll make you cry too. Probably even more.

"Savor every moment; they grow up so fast," say well-meaning strangers who've clearly suffered some closed-head trauma and now have amnesia.

It's true. Childhood doesn't last forever. But some days, it certainly feels like it will.

Special Shoelace Relativity

Time will cease to elapse

while waiting for the

five-year-old who has

just learned to tie her shoes,

to actually get the laces . . .

whoops . . . almost . . . nope . . . try again . . .

oooh, so close . . . give it another try . . .

okay, once more . . .

around and through . . . tied.

FINALLY.

School Field Trip Permission Form

Dear Parents,

As part of next week's unit lesson Respecting Reptiles, your child's class will be taking a field trip to THE WIDE WORLD OF GATOR ADVENTURES. It's sure to be a "wild" time! If you would like your child to participate in this fun-filled and potentially dangerous excursion, kindly fill out and sign this permission/indemnification form, have it notarized, and return it to your child's teacher by the end of the week. Please also enclose a check for $250. (Field trip fee includes transportation, park entry, lunch, snacks, and hazard insurance.)

Thank you!

Your School Staff

....

(Print Child's Name) ____________________ in (Print Child's Class) ________________ has my permission to participate in the class field trip to THE WIDE WORLD OF GATOR ADVENTURES on (Print Date) ______________, 20__.

Please provide the following information about your child:

Health issues: ____

____

☐ Does your child get car sick/bus sick?

(If so, we ask that you provide plenty of motion sickness bags. Parents whose children vomit on the bus and NOT in appropriate receptacles will be charged an additional $50 cleanup fee; make checks payable to the PTO.)

Food allergies: ☐ Peanuts ☐ Tree Nuts ☐ Dairy ☐ Gluten ☐ Soy ☐ Artificial sweeteners ☐ Food

Other (please list) ______________________________

____

____

Foods your child just doesn't like this week: ☐ Yogurt ☐ Mayonnaise ☐ Crusts ☐ Fruit ☐ Brown things ☐ Squishy things ☐ Slippery things

Other (please list) ______________________________

____

____

Additional Snack Release: My child has my permission to partake of school field trip snacks that are NOT (check all that apply) ☐ organic ☐ locally grown ☐ free of artificial colors or preservatives ☐ packaged in non-recyclable containers and are ☐ almost certainly devoid of any nutritive value whatsoever.

List the children your child is currently BFFs with: ______________________________

____

List the children your child is currently not speaking to: ______________________________

____

List all devices (with serial numbers) your child will be bringing on the field trip:

____

____

____

List the children your child is currently in a multi-player video game with: ____________

____

List person(s) to contact in case of emergency:____________________________

Your child's blood type: ______________________

Organ donor: ☐ Yes ☐ No

I understand that there is inherent risk to any activity, and I agree to hold the school harmless in the event that my child gets lost, left behind, injured, bitten or dismembered, or experiences death by gator as a direct result of the field trip, or simply finds the entire field trip "lame."

Parent Signature:____________________ Secondary Parent Signature:__________________

Witnessed By: ______________________

Signature of Notary Public: __________________

SEAL GOES HERE

School-Year Enthusiasm Trajectory

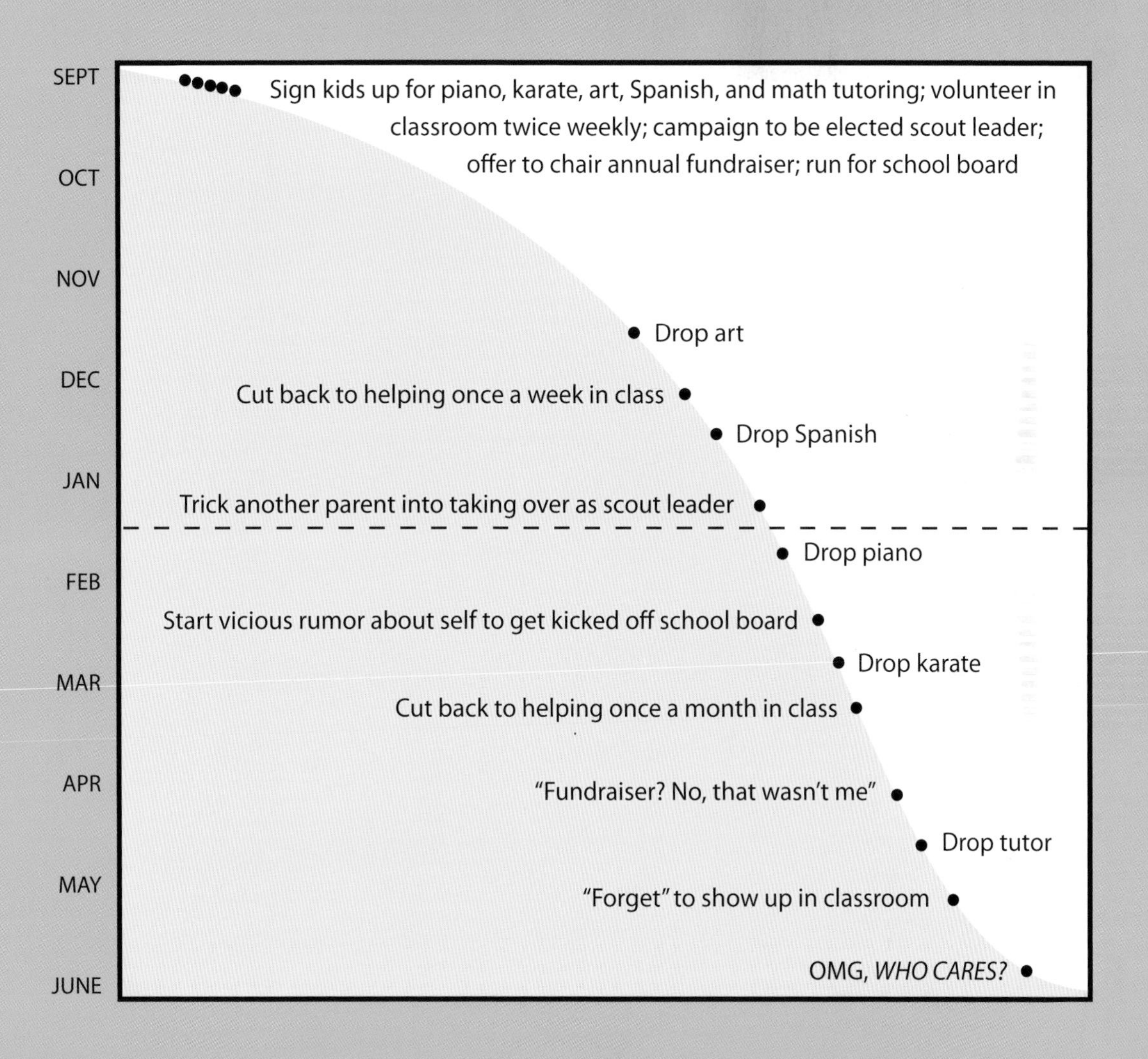

Quantum Carpool Mechanics

Why moms are able to drop off two kids . . .

at separate activities . . .

on opposite ends of town . . .

at exactly the same time.

Leaving the House with Your Toddler: The Five Stages of Grief

There is nothing as frustrating, as aggravating, as INFURIATING as trying to get a toddler dressed, fed, out the door, and (finally!) into her car seat in the morning. Attempt this feat while running very, very, very, late because you hit snooze a few dozen times before rolling out of bed, cursing, and every time your tot drags her little feet, you'll come perilously close to popping a blood vessel.

Although you may suspect your toddler is some kind of dawdling prodigy, dawdling, as early childhood development experts tell us, is a toddler's natural, preferred state. Dawdling is not meant to give you, the exasperated parent, grief. But it will.

Here, according to Dr. Elisabeth Kübler-Ross, are the five stages of grief that you will pass through in your attempt to leave the house with your child.

1. Denial

This is not happening. This is so not happening to me. Not today. Not today. Not today. I am NOT going to be late for my nine o'clock. Okay, okay, okay. Deep breath. Sweet Pea . . . Sweet Pea! Please let Mommy put your clothes and shoes on, 'cause we're a little late, okay, Honey? *I will not allow this to happen. This. Is. Not. Happening.*

2. Anger

Why does this shit always happen to me?! Stop playing with your toys and MOVE IT! *Every other mom gets her kid to daycare and gets to work on time. Why can't I?!? Why can't MY KID move her tiny little ass so I can get to work on time?* Just pick a pair of Crocs! The pink ones . . . I don't know where they are . . . the purple ones then. I don't know where they are either! Pick some shoes. I don't care! *Just ONCE I'd like to walk in without my boss glaring at me. She's got two kids; you'd think she'd understand.* I do not CARE that you don't like THIS princess T-shirt. You're wearing it! *But nooooo. She's got a nanny to get her kids ready so she can breeze in on time, looking all perfect in her Ann Taylor suits. Bitch. I bet she thinks I'm the worst mom ever. God, I hate her.*

3. Bargaining

Look, you can put on any clothes you want, okay? As long as you wear something, okay? I won't even brush your hair if you can just put some clothes on so Mommy can get to work, okay? Can you be a good girl and do that, for Mommy? *Please, dear God, if I can just get out of the house in the next ten minutes, I will never yell or swear again. I'll be a better mom . . . I'll give up*

wine ... I'll give up wine... for a day. Okay, a week. Two weeks! If I can just get her shoes on and leave in ten minutes, I promise I will give up wine for two weeks, okay?

4. Depression

This is hopeless. Why do I even bother? I'm never gonna make it on time. Maybe I should just call in sick. I feel sick. It's 8:20 a.m., and I'm exhausted. Maybe I'm getting the flu. Everything just sucks. My life sucks. My job sucks. Motherhood sucks. I suck. I'm such a loser. I need wine.

5. Acceptance

Look at you. You're all dressed. You look, um—*Good God. My child looks like a colorblind crazy person dressed her. Cheetah print with plaid? Whatever. She's wearing clothes. Focus on that. Clothes are good. Deep breath.* You look beautiful, Sweet Pea. *Another deep breath. We might actually make it to the car.* Can you step into your shoes? Thatsa girl. One foot. The other foot. There we go. *It's okay. It might actually be okay.* Are you ready to get in the car? *And we're walking, we're walking... I'm actually opening the car door... she's getting in the car... I'm buckling her in ... I'm closing the door ... It's going to be okay. I'll be a few minutes late, but I'll make it. I'll just tell 'em my watch needs a new battery, and it's running slow.* What's that, Sweet Pea? You have to go potty? Are you kidding me? You need to POOP?!? Of course you do.

Destination Arrival Equation

% charge left
on electronic devices

Distraction Quotient
videos available • pairs of working headphones

$$(CL \times PS)DQ^2/BS = X$$

Personal Space
arguments

Bathroom stops
$\left(\frac{\text{actual stops}}{\text{planned stops}}\right)$

Miles to go

Heisenberg's Theme Park Uncertainty Principle

The more certain you are of your child's

precise location,

the less you can know

about his speed and direction when

the impulse strikes, and

he takes off running.

Conversely, the more you know

about your kid's speed,

the less likely you are to know

where the hell he actually is.

EVERY PARENT NEEDS A FALLOUT SHELTER

Early on in this experiment called Parenthood—right about the time tots find their legs and take off running—parents notice that small children have boundless energy. Which is actually like some cruel cosmic joke. Because as a parent—ha-ha!—you have none. Zip. Zero. Nada. Bupkis. This also explains why parents throw back venti coffees spiked with 5-Hour Energy just to get through story time.

No doubt, more than one parent has pondered the possibility of winning a Nobel Prize *and* liberating the country from dependency on fossil fuels, if they could just figure out how to harness that endless energy source.

Ohhh, but that particular energy source is a highly volatile one. Physicists—certainly those with children—describe it as "fissionable material," the handy-dandy stuff used to fuel atom and hydrogen bombs. Which, when you're deep into the Terrible Twos, with their hair-trigger tantrums (or the hormonally driven tweens with their back talk and attitude), seems just about right.

Living with an explosive child will make your home feel like a war zone as conflicts continually erupt over naps, homework, not licking the window sill, and whether today's sippy cup should be blue or green. Still, there's always hope for détente.

Morning Energy Equation

$$(E = MC^2)$$

$$\text{Energy} = \text{Mom} \times \text{Caffeine}^2$$

Views of the World

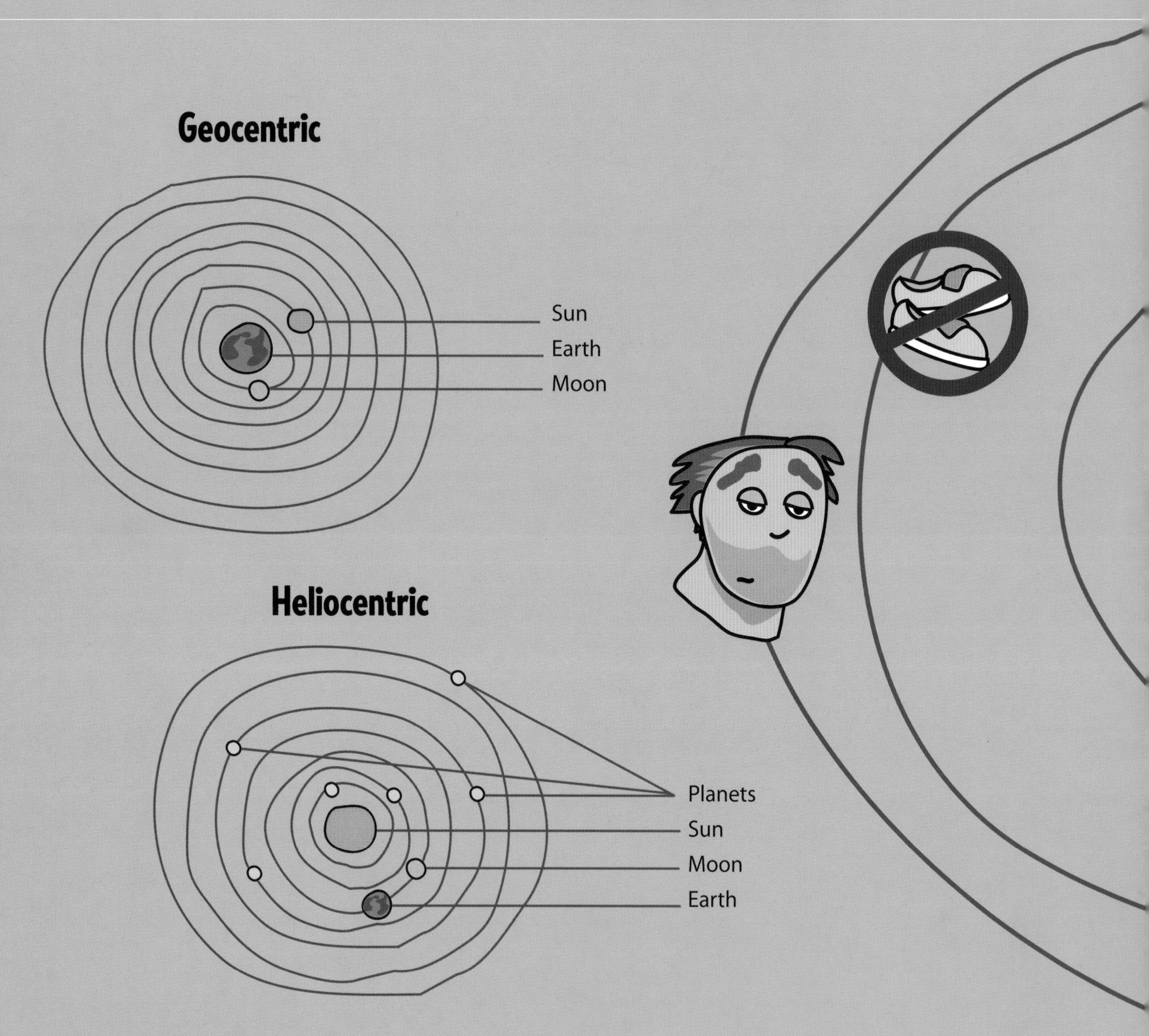

Toddlocentric
MINE!
NO!
I DO IT!

The Physics of Getting a Child Out of Bed

Atom blasters and particle detectors are great gizmos to have on hand should you need them. But sometimes a simple device—like the lever and fulcrum—is really all that's required to get the job done. This combo has been used to pry apart and lift heavy loads with minimal force at least since the Egyptians went on their pyramid-building spree. Today, parents can use this same simple technique to dislodge kids burrowed deep in the blankets, ignoring all demands to GET OUT OF BED RIGHT THIS MINUTE! We can thank the ancient Greek scientist Archimedes for figuring out exactly how the lever-and-fulcrum works, no doubt when he was trying to pry *his* kid out of bed in the morning.

Technique for Getting a Child Out of Bed on a School Day

1. Open window blinds to flood the child's room with sunlight.
2. Nuzzle the child softly. "Good morning, Sweet Pea. It's time to get up now."
3. Gently jostle the child's shoulder, tousle the hair, lightly tickle the toes and neck. "C'mon, Sweetie Pie. Time to get up. Let's go. We don't want to be late for school."
4. Stand over child, giving the shoulder a slightly more vigorous shake. "C'mon! We're going to be late if you don't get up. Breakfast is on the table. NOW, please!"
5. "GET UP! NOW! **DO NOT** MAKE ME TELL YOU AGAIN!"
6. This is the point when the lever-fulcrum is most useful. When you observe that the child has curled into a tight ball under the covers, pull back the blanket, slide your arms beneath the child, bend the knees, rock back, and roll that kid out of bed and up into your arms. Set the child down at the kitchen table for breakfast.

Congratulations! You have successfully removed your child from bed.

Technique for Getting a Child Out of Bed on a Weekend

1. Allow the first rays of the sunrise to brighten the child's room, and ..."I'm wake! Let's play! Get up! I'm hunnnnnngryyyyyy! Get up! Get up! Get up! Get up! NOW!"

Coefficient of Friction

The resistance encountered
when dragging a
shrieking toddler
out of the park
at naptime.

Contents Will Explode (It's Only a Matter of Time)

When physicists talk about devastating explosions, they typically list the neutron, atom, and hydrogen bombs as their top picks.

They are wrong.

Using the Parental Freak-Out Scale, few things rival the Toddler Tantrum for shock, awe, and damage (at least to the parental ego).

Triggered by seemingly random, utterly innocuous events—stirring chocolate powder into milk; reading the "wrong" story; not reading it enough times—the Toddler Tantrum is designed to explode without warning and bombard its target with screams, kicks, tears, slobber, and—if a parent is crazy enough to jump in and try to contain it—elbows to the eye and knees to the groin. Among scientists and Defense Department staff, that is known as "collateral damage."

Toddler Tantrums are terrifying in the privacy of your own home, but it's the public outbursts—typically set off in supermarkets, restaurants, or your in-laws' antique-filled homes—that have the most fallout, as parental judgment rains down. Trauma specialists believe that the judgiest parents, tsk-tsking, MY child never behaves like "THAT," are actually suffering from Pinocchio Syndrome, a fugue state that allows them to block out the most embarrassing things their children have done as a coping mechanism. (But feel free to throw withering glances their way. Just because they're in denial doesn't mean they have to be asshats.)

Toddler Tantrums are as inevitable as death, taxes, and vomit down your back. But to quote Edna Mode of *The Incredibles*, "Luck favors the prepared, Dahling." Here's how to survive when your tot goes all Fat Man and Little Boy.

Remember your scouts training.

Boy Scouts, Girl Scouts, Cub Scouts, Indian Guides—everyone was some kind of scout, so be prepared. This is not like the Cold War, where each side had an effective deterrent against shooting nukes at each other. There is no known deterrent for a Toddler Tantrum. So be sure you have a bunker available. In a pinch, any interior room (closet, laundry room, bathroom) will do, provided it is stocked with construction-grade earplugs, vodka, and paper bags for hyperventilation. And cookies. You were a scout. Remember the cookies.

Run.

Recognize the signs of impending tantrum before it explodes and get as far from the blast source as possible.

Duck and cover.

Seriously, you do not want to get kicked in the head or punched in the solar plexus.

Let the tantrum run its course.

It's tempting to try to contain the explosion. Rookie mistake. That's when most parent injuries occur, as flailing tots are freakishly strong and can easily deliver a swift roundhouse to the face. (Pro tip: keep ice packs in the freezer. And a cosmetic surgeon in your favorites list.)

Eat the cookies. Chase with vodka.

Assess the damage.

If the tantrum was an in-home event, simply make sure nothing broke, like your mother-in-law's favorite Tiffany lamp or your tot's hand. If, however, the explosion occurred in public—especially in an enclosed space, like a 747 or a quiet French restaurant—you're on the hook for a round of cocktails for everyone, even the pilot and maître d'. And if you took shelter in the bathroom while your tot screamed like she was being filleted with long knives, collect everyone's bank-routing numbers so that you can make direct contributions to their IRAs. Just think of it as reparations.

After-School Fission Effect

Just when you think you know what you're doing, that you've gotten into a rhythm with this whole parenting thing, along comes homework to blow it apart again. All that confidence goes flying off again, like so many spent atoms.

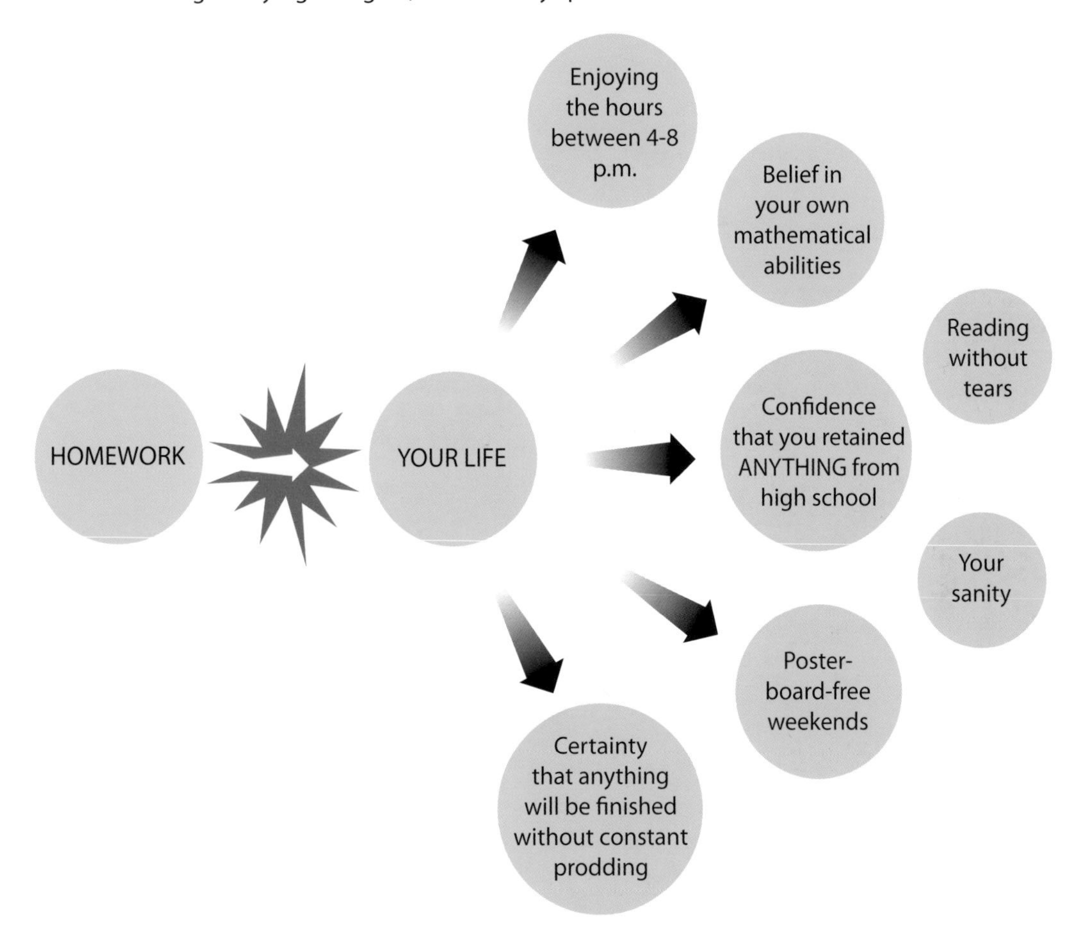

Faraday's Rage Cage

Mom's well-shielded
and highly fortified safe space
for waiting out the Terrible Twos.

And possibly the Threes.

"A mathematician is a blind man in a dark room looking for a black cat, which isn't there."

- Charles Darwin
Naturalist/Geologist

Mathematics

SUPPLY-SIDE ECONOMICS

For most parents, the concept of supply-side economics is a fairly intuitive one: We supply it. The kiddos spend it—at least until we get into deficit spending. At which point the bankruptcy administrator usually takes away the credit cards.

No question, kids are spendy little buggers. Between feeding them, clothing them, and keeping them in American Doll accessories, parents can easily pony up the equivalent of a small nation's GDP to shepherd a kid from mom's womb to dorm room. And that's before the *please-please-please* pleas for real ponies start rolling in.

Even assuming you nix the pony, there is always something to buy a child. If you've ever wondered why moms and dads drive beater cars, it's because their discretionary income is being slowly siphoned off to fund extracurricular activities, umpteen zillion school fundraisers, and at least a decade's worth of crappy plastic goodie bag fillers that the receiver's mom is going to deposit in the garbage as soon as her kid's back is turned.

Of course, parents don't begrudge the money they lay out. Even if it means eating cold cereal every night to pay off a fancy new piano so Little Mozart can spend three minutes a week playing "Chopsticks." But really, without that long-term investment, how else would our kids get into a college we can boast about on our car's back windshield?

Now, if only universities would accept tuition payments in Lego pieces.

Baby Registry Essentials

The *concept* of the baby registry is an economically sound one: Babies cost a crap-ton squared to raise, so parents-to-be rope their mothers, fathers, and anyone they've ever known to chip in so the kiddo has everything he might possibly need to prevent him from growing up to be a sociopath. Really, it's the ultimate Kickstarter campaign.

However, the *problem* with baby registries is that no one has any clue about what they're *really* going to need until that baby is well into middle school. And by then it's a bit late to request a Diaper Genie.

Alas, it's all too easy for expectant parents to get dazzled and distracted by "totes adorbs" crib bedding with matching lamp shades, mobiles, and valances while wandering the aisles at Buy All The Things Baby Emporium. That's the kind of stuff parents want *before* the baby is born. Here's what parents *wish they had* after that baby comes home:

- Crib
- Stroller
- Car seat
- Basic car seat installation course
- Advanced car seat installation course
- Gift certificate for professional car seat installation
- First-month starter pallet of diapers
- Cry-to-English translation dictionary
- Three-hundred pacifiers in assorted colors and shapes
- Get-out-of-breastfeeding-free card, entitling bearer to one guilt-free decision to switch to formula
- Bedside taser to nudge your spouse awake for 4 a.m. feedings
- His & Hers $1,000 Starbucks gift cards
- Panic room, equipped with Godiva, Jim Beam, and a child-proof door knob
- Noise-cancelling headphones
- Protective sheeting for kitchen walls, windows, and doors for when baby starts solids
- Divining rod for determining which food the child might eat on any given day
- Twenty thousand boxes of Band Aids
- Tax-free savings plan for replacing lost pacifiers, blankets, lovies, thermoses, jackets, socks, shoes, gloves, sweaters
- Find-My-Former-Self App to locate Mom's pre-baby abs, boobs, brain cells, and bladder control
- Fifty-two prepaid Saturday nights with a grandmotherly babysitter who raised eight kids and is fazed by nothing
- Thirty-day supply of Klonopin . . . with eleven refills

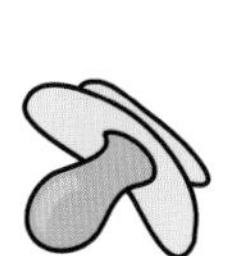

Fuzzy Logic

Buying your toddler's favorite stuffed lovey

by the gross

so you are never caught

without a spare.

Orders of Magni'tude

A fool and her money are soon parted: In an effort to teach a modicum of financial responsibility, a parent making even the most benign suggestions about how to manage money will be met with a level of attitude many times greater than is really necessary.

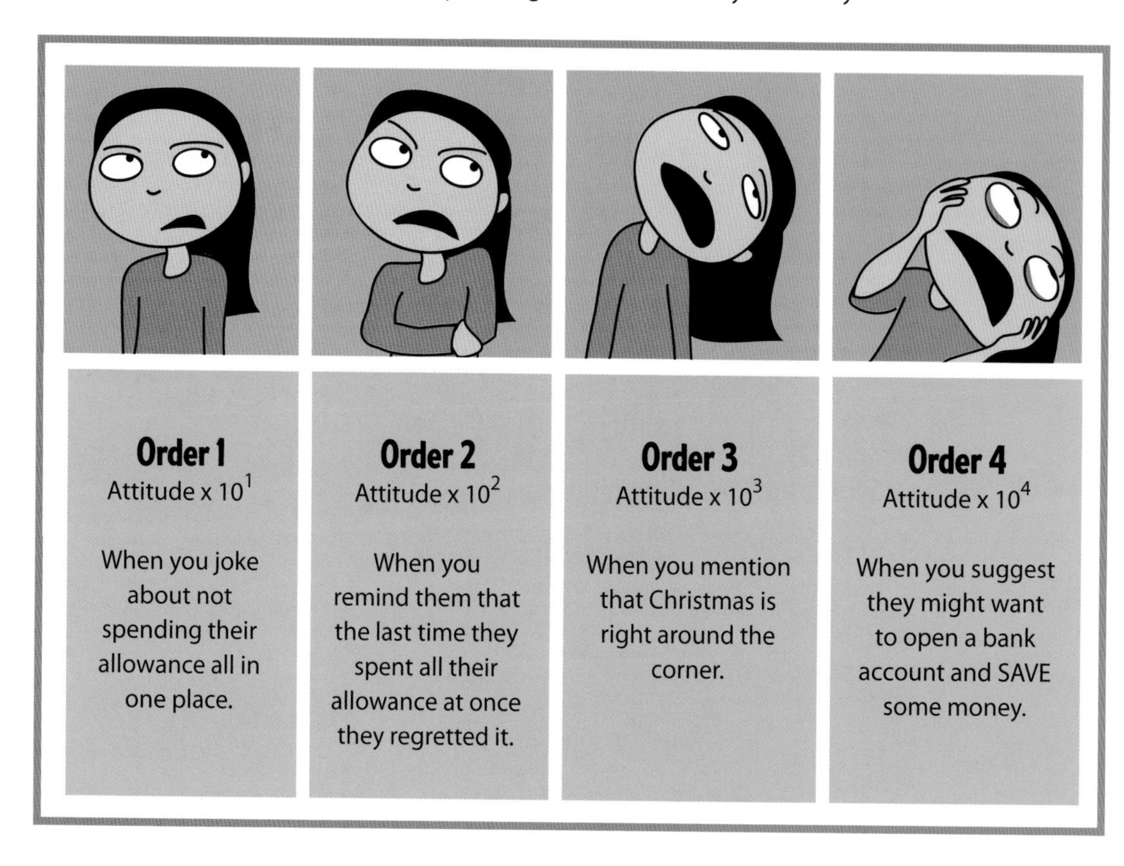

Play-Doh to PlayStation: Where ARE Your 401(k) Dollars Going?

Perhaps you saw the report that came out not too long ago from the U.S. Department of Agriculture. The one that estimated it will cost the average parent somewhere in the neighborhood of $305,000 to raise a child born in 2013 to age eighteen.

That's $305,000. Think about that for a moment. Really let that number sink in. We'll let you catch your breath as you wonder where the hell that money will come from. Hyperventilating yet? Try not to; paper bags cost extra.

So . . . now that your blood pressure is somewhere in the exosphere, let's ponder the implications. Apart from confirming our suspicions that children actually *are* little wild animals that need to be domesticated before they take their place in polite society—why else would the *Department of Agriculture* issue that report?—here's what really struck us about the USDA's announcement:

1) At least a third of that figure is replacing lost *Frozen* and *Star Wars Rebels* thermoses
2) $305,000 only gets us halfway to adulthood

Seriously, what about the REALLY BIG expenses like college, graduate school, the vineyard you'll need to keep you in fermented grape juice as you mortgage your soul to fund the parenting process?

What sticker-shocked parents could really use is a metric to determine their personal spending style and how financially prepared they are to shepherd their children from birth to self-sufficiency—which if you go by the milestone charts, typically happens around age forty-nine or fifty, give or take. And when it comes to children, let's be real, it's mostly take.

Fortunately, we have such a metric. Working closely with Blood From a Stone Financial Services and the developers of the *Cosmo Quiz*, we designed the Science of Parenthood Financial Readiness Scale—seven simple questions that can gauge parents' fiscal preparedness, as well as which Kardashian they resemble most.

Remember, there are no incorrect answers here. (But if you have money left over, you're doing it wrong.)

1. Which fictional character most closely describes your financial style?

A) Scrooge, but not as spendy.

B) I'm like the ant from the Aesop fable. Slaving away to save, save, save.

C) Style? What, like Louboutin versus Converse?

2. Who directs your household budget?

A) I do.

B) My spouse.

C) We make financial decisions together.

3. Cut the crap. Who REALLY dictates family spending?

A) My mother-in-law.

B) The kid.

C) The bankruptcy administrator.

4. Where do you keep your money?

A) Diverse portfolio of mutual funds managed by a blue-chip investment firm.

B) Money market account at my local bank.

C) I like to keep close tabs on our money, so it's in cars, gaming systems, Rolex watches, designer purses, and shoes.

5. What's your biggest monthly expense?

A) Mortgage/rent and groceries.

B) Pacifiers, diapers, formula.

C) Beanie Boos (These will be worth a FORTUNE someday!).

6. How do you calculate a "worthwhile" expense?

A) Can we afford it? Is there money in the budget?

B) Price of the item *plus* time spent arguing with the kid about it *plus* time the kid will whine about it *minus* my ability to tune out the whining *minus* how quickly I can pour myself a cocktail *equals* total real cost.

C) I imagine the joy it will give my children . . . or whether I really, really, really want it. That justifies any expense.

7. Where do you see yourself when your nest is empty?

A) Empty? We built an addition on the house for the kids to move into.

B) In a single-room occupancy unit, hopefully with heat.

C) Witness protection.

QUIZ KEY

If you chose mostly A's: Great job! With your careful planning, you could have a Brangelina-sized family and not feel the pinch. Ponies for everybody!

If you chose mostly B's: We feel for you, we really do. Your kids are bleeding you dry. Hopefully once they reach eighteen, they can start supporting you.

If you chose mostly C's: Unless you've got money stashed in a numbered Swiss account, head to Vegas and put what's left in your pockets on black. *Bonne chance!*

Wintertime Scatter Graph

Like dead leaves scattered by winter winds, so too are kids' pricey sweaters, jackets, hats and mittens dropped, left, lost, and forgotten all over town.

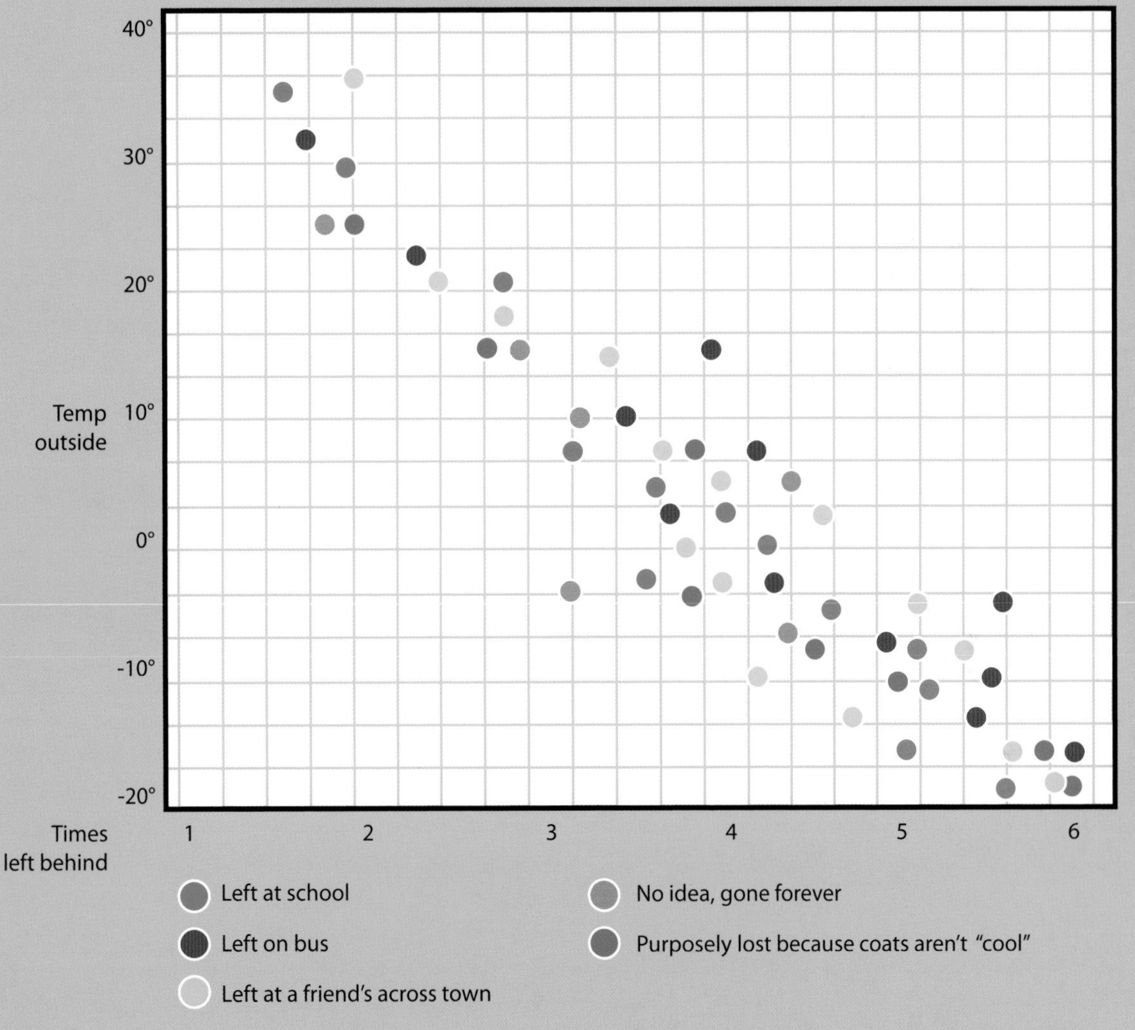

Occam's Stroller

When choosing between two models,
simpler is better . . .
which will
become obvious after
you've bought
the fancier ride.

Chuck E-conomics

The complex formula
that determines the exact
dollar-to-arcade-token
exchange rate . . .
allowing you to calculate how much
real money you've shelled out
for your kid to win
a $1 prize.

GAME THEORY

There comes a time in every child's life, not long after she learns to purposefully close her little fingers around an object and hold it freakishly tight, when she begins to grasp the socio-economic concept behind smash-and-grab capitalism. As in:

> If I'm holding it, it's mine.
> If I snatched it from you, it's mine.
> If I found it on the ground, it's mine.
> If I put it in my pocket, it's mine.
> If I looked at it, it's mine.
> If I momentarily thought about it, it's mine.
> If it's at a store, it'll eventually be mine, so basically, it's all MINE! MINE! MINE!

Some folks call this greedy. Toddlers call this logic. Still others call it a lifestyle choice. (The latter are also called "criminals" and usually end up doing two to seven in a state penitentiary.)

Fortunately, when exposed to enough *Sesame Street*, most kids eventually catch on that they can't have everything they want... though typically not until we've been worn down by torrential begging and then find the object of their desire lying discarded on the floor.

To quote '80s icon and toddler role model Gordon Gekko, "Greed is good." Better still would be a "gently used" return policy.

Should You Ditch That Toy?

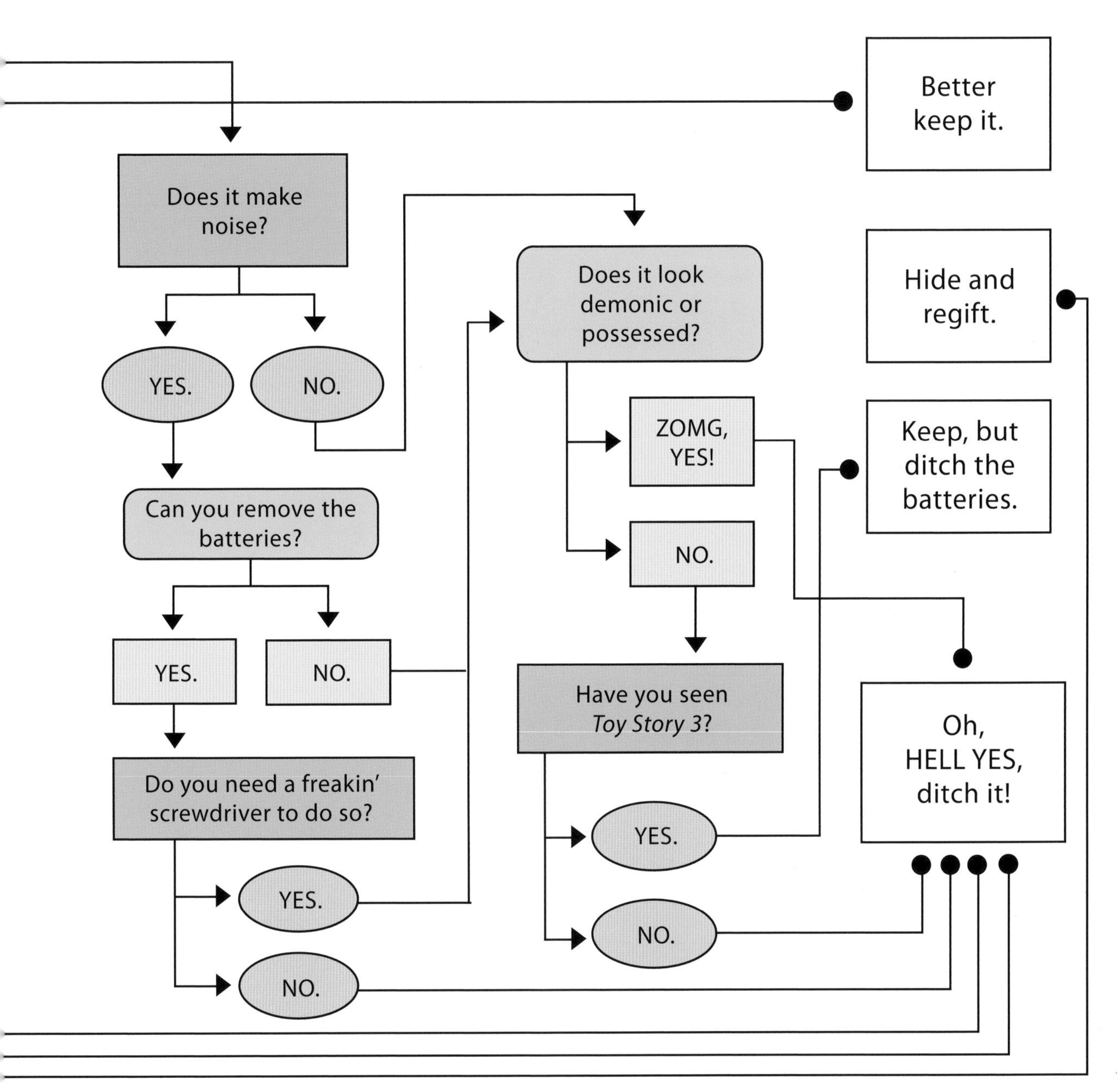
Better keep it.
Does it make noise?
YES.
NO.
Can you remove the batteries?
YES.
NO.
Do you need a freakin' screwdriver to do so?
YES.
NO.
Does it look demonic or possessed?
ZOMG, YES!
NO.
Have you seen *Toy Story 3*?
YES.
NO.
Hide and regift.
Keep, but ditch the batteries.
Oh, HELL YES, ditch it!

Newton's First Law of Parenting

A child at rest will remain
at rest . . .
until you need your
iPad back.

The Toy Pyramid

Ask about the best toys for children, and any Vegas bookie worth his pinkie ring would lay odds on toys that keep a kid quiet, out of traffic, and away from the lighter fluid and matches. Those same playthings are sure to receive high marks from frazzled parents looking for fifteen minutes of peace between work and dinnertime to pay bills, answer email, and plan that enchanting Instagram-ready family vacation to be featured on next year's holiday cards. And that's why video game systems and tablets are so beloved.

Surprisingly, though, some in the parenting trenches believe toys should do more than distract kids from otherwise playing chicken on their bikes with the neighbors' cars—even if they are wearing helmets. That is why the ad-hoc cooperative group Families for Unplugging Captive Kids from Internet Technology devised a new Toy Pyramid as part of its "get-the-hell-back-to-basics" play initiative.

Based on the food guide pyramid, the Toy Pyramid is aimed at peeling kids from the small screens and plugging them back into three-dimensional life. It eschews today's digital shoot-the-zombie games in favor of PPE toys—toys that Predate the Pong Era. No word yet on how moms and dads are supposed to get anything done or form a coherent thought with all these kids running around, playing.

New Toy-Enthusiasm Curve

When a child receives a new toy, his or her interest in that item will follow a predictable downward trajectory and then come to rest at an established baseline, unless it is influenced by an outside force.

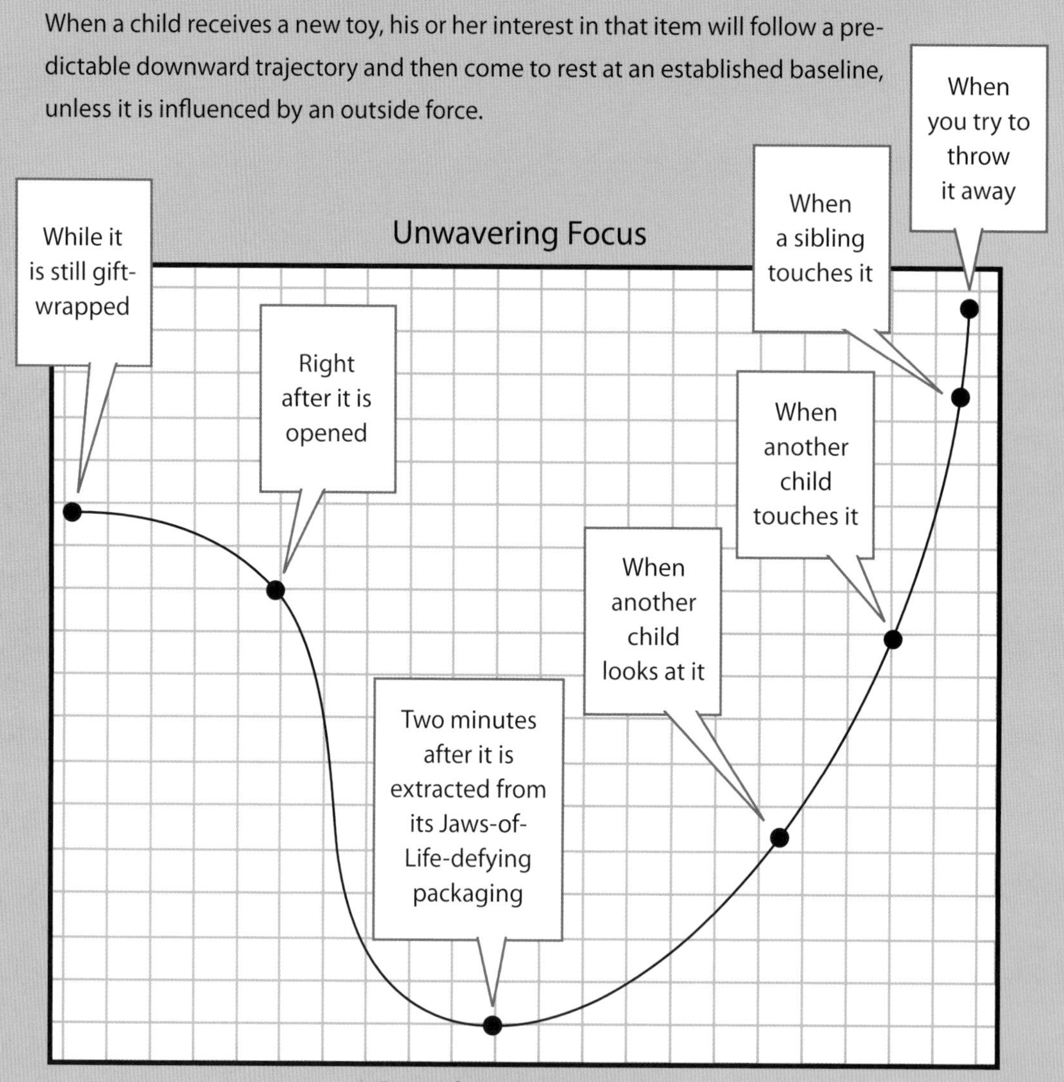

Shut Your Pi Hole!

Like pi—3.141592653589793238462643, ad infinitum—children do tend to chatter on and on and on . . . forever. They'll prattle endlessly about the benefits of this Nerf blaster relative to that Nerf blaster or what some kid you've never heard of did at school or the plot points of a *Gravity Falls* episode you've never seen.

Naturally you want to know what's going on in your children's lives, who they're friends with, what TV shows they like, and their latest brilliant excuse for not picking up their toys. Sometimes you can listen with rapt attention, marveling at how their little minds work. And sometimes you want them to just shut the hell up. Alas, children remain stubbornly unaware that silence is golden, and that some days, like when Mommy has a headache (or a hangover), SILENCE, DAMMIT! is more valuable than a diamond mine in Minecraft.

Still, you don't want to scream (especially with a hangover). Before you turn your car into oncoming traffic with the fervent hope that at least in a coma you might finally get some peace, try playing this fun parenting game.

1. You're stuck in heavy traffic on a long road trip. After the thousandth "Honk the horn, Mommy! Just go, Mommy! Why aren't we moving, Mommy! GO!" You . . .

A) tell the kids it's time to play the Quiet Game, promising a pony if they can keep their traps shut until you get there.

B) turn up the heavy metal station so you can't hear them whine over Megadeth.

C) chant "serenity now . . . serenity now . . . serenity now" while rocking back and forth in your seat.

D) employ your secret stash of duct tape.

2. The kids have been singing "Dumb Ways To Die" for the last hour, and the song is driving you crazy. You . . .

A) quickly pop in earplugs from your handy hundred-pack.

B) put on "It's a Small World," hoping two intensely annoying songs will cancel each other out.

C) sign them up for singing lessons.

D) reach for the duct tape.

3. You've told your three-year-old repeatedly that you're not buying plastic dinosaurs at the store today, but she's maintained a steady "Please, Mommy! Please, Mommy! Please, please, please, Mommy!" as you've wheeled your cart from one end of the store to the other. You . . .

A) buy the damn dinosaurs. It's worth $10 of cheap plastic so you can hear yourself think.

B) tune her out and focus on your list. Your three-year-old is not the boss of you.

C) threaten to put the kid in time-out till she's twenty if she doesn't STFU.

D) hit the home-improvements aisle to grab a fresh roll of duct tape.

4. Your kid has been weaving a fantastical tale that combines elements of *The Fox and the Hound*, *Dr. Who* and *Star Trek* for the last forty minutes. You . . .

A) applaud your future Booker Prize winner for her creativity.

B) half listen while wondering, *How much longer till I can turn on NPR?*

C) call out, "Look at the squirrel!" hoping to distract her.

D) quickly tug on a roll of duct tape—*zipppp!* That ought to do it.

5. That basic bitch Cindy from Baby Aquatics is bragging (again) that her two-year-old made the nursery school honor roll, speaks fluent Mandarin, reads on an eighth-grade level, and has been invited to perform a violin solo with the Boston Pops. You . . .

A) fake appropriately impressed sounds while scanning the class schedule for a different time slot.

B) fake a medical emergency so you can duck out of the convo.

C) tune her out by doing complex math in your head.

D) slap her collagen-filled lips with six inches of high-tack duct tape and beat feet. She'll never know what hit her!

QUIZ KEY

The answer is always duct tape.

Newton' s Third Law of (E)motion

Your gift will be met with a

measure of disappointment

that is equal to the certainty that

"she will absolutely love this!"

with which you bought it.

EQUATIONS FOR INSANITY

When Malcolm X quipped that "mathematics leaves no room for argument," he'd obviously forgotten what it's like to reason with a sobbing four-year-old that two halves of a hamburger DO equal a whole, and that the burger will taste good even if it fell apart a little on the plate.

Math is supposed to be pure. Absolute. We rely on numbers to offer explanations, even for the insanely unexplainable. Like why deeply linked particles affect each other even when separated by great distances—which parents of twins no doubt totally understand.

Numbers, for the most part, do a pretty good job. Even that weird particle thing is explained by quantum entanglement and the speed of wave function collapse. Not that *we* could calculate that. Still, it's reassuring to know that the math accounts for the odd behavior.

But good luck applying mathematical reasoning—hell, *any* reasoning—to kids' behavior:

- How does a kid eat a PBJ every day but gets bored waiting for an app to download?
- Why are kids attracted to swings only if there is a forty-five-minute wait to use them?
- How can a kid gnaw through a binky's medical-grade silicone but can't chew and swallow boiled chicken?

Alas, like that other mathematical mystery involving the number of licks and the radius of a Tootsie Pop, the world may never know.

The Mobius Strip

The never-ending process
of getting your child
dressed
for the day.

Irrational Numbers

Mathematicians define irrational numbers as "numbers that are not rational," which pretty much accounts for just about every number that comes out of a child's mouth. As any parent who's ever asked how many minutes their kid's eyeballs have been glued to a screen knows, kids have an *unrealistic* sense of how things are quantified. The same kid who swears that he's read for "twenty hours" will cop to just five of the sixty minutes of video games he's played while Mom was busy on a conference call.

Some more numbers that kids get irrational about:

How long are you going to be gone? **TWO HOURS?!?** That's FOREVER!

I took three bites! That's A LOT!

Mommy, how old are you? Are you a hundred and fifty, Mommy? Were there dinosaurs when you were born, Mommy?

Where were you?!? I thought you forgot me! I've been waiting for you for A MILLION HOURS!

Why can't I get the Power Wheels Cadillac!? It's not a lot, it's only $420.

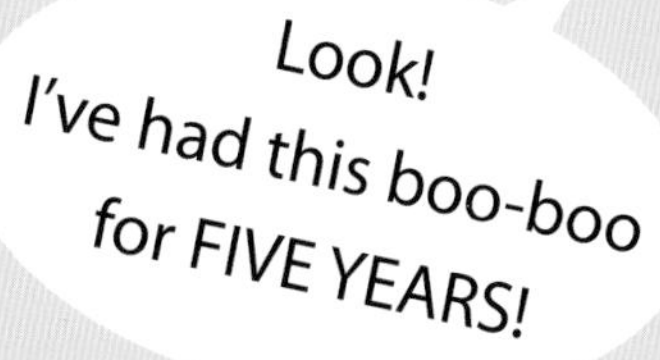

Grandma sent me five dollars! I'm rich!

Fractional Distillation

The disheartening revelation that
no matter how it's broken down,
you still cannot comprehend
your child's "new" math
homework.

Solid Geometry

"Studying"
the shape, length, and circumference
of the poop
your child insists you see
before he flushes.

Parent-Child Communication

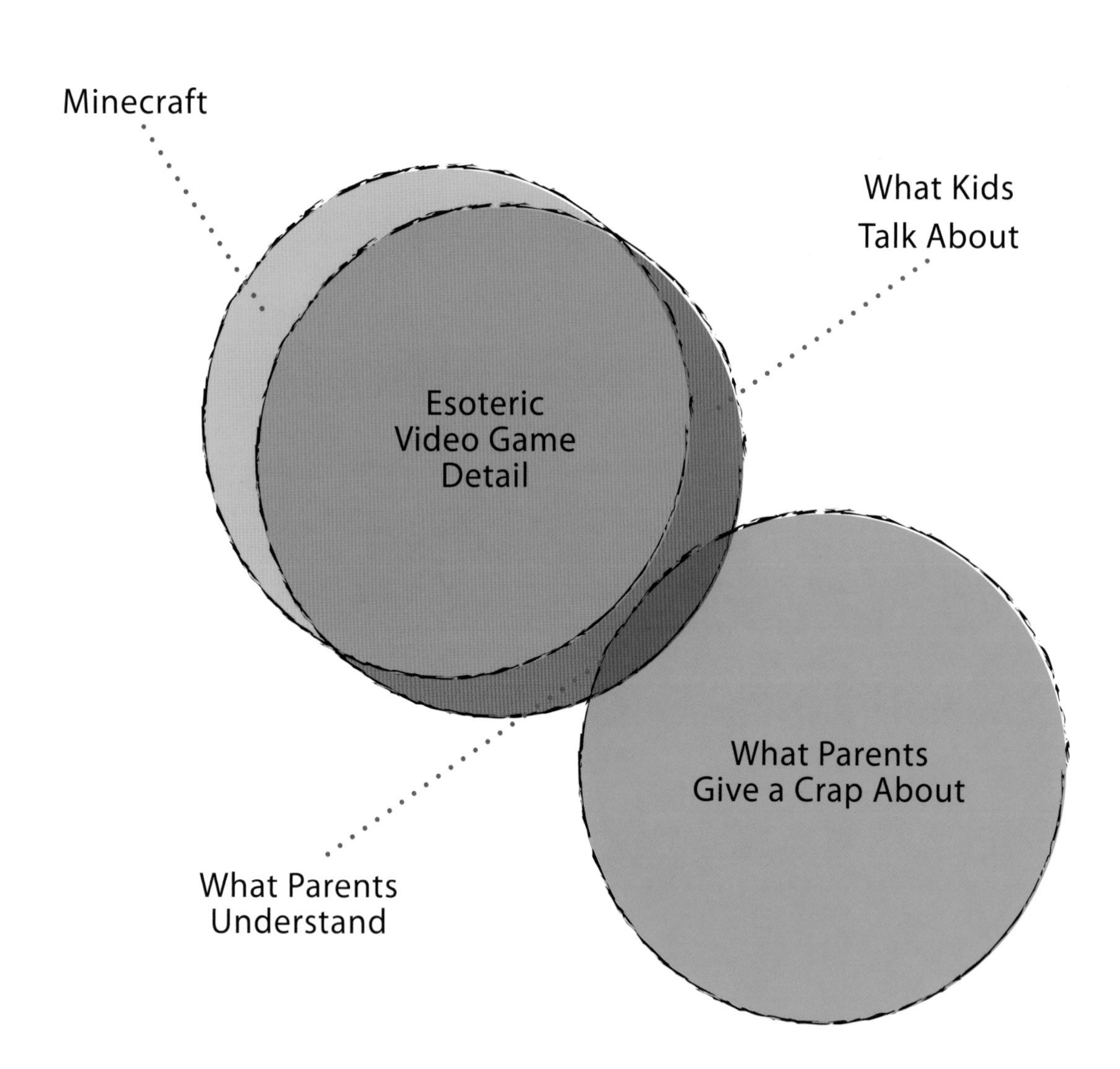

Applied Parental Mathematics

Surveys show that approximately 99.9 percent of American teenagers believe math is a complete waste of time that would be better spent SnapChatting with friends, and that once the SATs are d-o-n-e, they can kiss those crazy word problems goodbye.

If only that were true. Higher-level math starts up again as soon as moms clear recovery, as they immediately start calculating the ratio of breast milk to wet diapers and what portion of their C-section is covered by insurance. The good news is that, according to anecdotal evidence, there is something about the act of becoming a parent that turns even complete math idiots into scholars.

Scientists haven't quite teased out how that works, but the effect holds whether parents adopted, used egg donors and gestational surrogates, or got pregnant via a good old-fashioned drunken weekend in Vegas. The takeaway is that once you have kids, challenging real-world word problems like these are a snap. (Though, sadly, Common Core math homework will not get any easier.)

1. Two-year-old Marcus's favorite bedtime story is *Trucks Roll!* If the book has 147 words in it and Marcus's mom Joanne reads it at least four times each night before bed, how long before she pretends the dog shredded it so she can read any other goddam story?

2. For her son's seventh birthday, Monica rents a bounce house, a rock-climbing wall, and two ponies. If Monica sends out twenty-five invitations and twenty parents RSVP that their kids will be there, how many will actually show up?

A) Calculate the number of kids who will bring an uninvited sibling ("You don't mind one more, right?")
B) Calculate how many children will ignore the bounce house, etc. to play Minecraft, indoors, on their own devices.
C) Calculate how long after the bounce house, climbing wall, and ponies are taken away that the birthday boy will wail, "It's not fair! I didn't get a chance to do anything! This is the worst birthday ever!"

3. Lindsey's first-grader needs to read for twenty minutes every night. If her first-grader starts at 7 p.m. and spends ten minutes staring blankly at the bookshelf "picking a book"; ten minutes listening to all of the ringtones on the iPad before choosing one for the timer; fifteen minutes whining that the book is too hard . . . too boring . . . too heavy . . . too long; and then ten minutes bargaining for a video game break, at what point will Lindsey demand that her husband "deal with this," then hide in her closet binge-watching *House of Cards*?

4. Brenda purchases two first-row orchestra tickets to take her three-year-old to the stage show *Dinosaurs Walking*. The tickets cost $75 apiece, and it takes her forty-five minutes to drive to the theatre and costs $10 to park. Determine how long after the show starts that her three-year-old will start crying, "Scary, Mommy! I want to go home! Go home NOW!"

A) How many times does Brenda explain that if they go home, they can't come back? Round to the nearest ten.
B) How long after Brenda pulls back into her driveway does the three-year-old start crying to go back and see the rest of the show?

5. Sammy spies a Super Soaker water gun that costs $19.99 at Target. He chatters nonstop about the Super Soaker for thirty minutes, promising to "never ask for anything else, ever, ever, ever again" if he can have it. If his mom finally agrees to buy the water gun, estimate, to the nearest minute, the time it will take Sammy to lose all interest in the Super Soaker once he gets his new toy home.

Extra credit: When Sammy's younger brother picks up the forgotten water gun, estimate how many seconds it will take for Sammy to snatch it back, yelling, "Don't touch it! That's mine! You can't have it. MOM!!!"

6. Maggie has eight popsicles in assorted flavors. If she divides them equally between four children, how many kids will pout that they don't like the color popsicle they got?

Extra credit: Calculate how much Maggie actually cares.

7. Janie begs her parents every day to "Please! Please! Please!" take her to ride a roller coaster called The Beast. If her parents give in to the endless begging and agree to drive two hours to the amusement park, pay $89.99 per ticket to enter the park, and then wait in line for ninety minutes, by how many fractions of an inch will Janie be too short to ride?

Extra credit: How many overpriced park souvenirs will Mom and Dad need to purchase before Janie stops sobbing?

8. Rebecca and Connor take their four children on a road trip from Baltimore to San Jose. If they leave their house at 10 a.m. on a Monday, driving 65 miles per hour, how many times will they have heard "Stop breathing on me!" "Mom! He's looking at me!" "Am not!" "Are too!" "Cut it out!" "MOM!" before they cross the Maryland state line?

A) How many days into the trip before Rebecca starts adding a shot of vodka to her morning OJ?
B) How many shots will Rebecca have added by the time they reach Missouri?
C) True or False: There is no longer any OJ in Rebecca's "juice."

Parental Social-Life Equation

BIRTHDAY PARTIES
(offspring x classmates) % likelihood of invite

weekends per year

average illness rate

(BP + HS / WpY).25 = X

HOSTAGE SITUATIONS
offspring (team sports + activities + scouting events + awards ceremonies)

chance you will ever have an adult social life again

Sleep Geometry Theorem

A child will always sleep

perpendicular

to any adult(s)

sleeping next to him.

"Anyone who thinks science is trying to make human life easier or more pleasant is utterly mistaken."

- Albert Einstein
Physicist

Honors and Accolades

Science of Parenthood's Eureka! Awards

Honoring Major Parenting Moments That Totally Deserve a Prize

Just as the Fields Medal celebrates outstanding discoveries in mathematics, and the Nobel Prize honors advances in the sciences, Science of Parenthood's Eureka! Awards recognize breakthrough moments in parenting and family life. So few are given because these moments . . . they are so rare.

And the Eureka! Award goes to . . .

Mathematicians **Hartwin Mansfleck** and **K. Willis Treacle** at the University of Chicago for determining the numeric value of "crush depth"—the precise point at which relentless whining for the latest Xbox system, iWhatever, or Fashion Barbie's Manolo Blahniks will break even the most stalwart parent's resolve, causing it to crumble like a stale graham cracker.

Microbiologists **Drs. Reginald P. Smoot** and **Farley McWarblings** at the University of Oxford for their groundbreaking research on Lego proliferation. Legos' propensity to multiply once they enter a home is well-documented, but it was the Oxford team of scientists who discovered that Legos' replication mechanism was nearly identical to the way viruses reproduce and spread. They noted in their research paper that "just as a child's cold virus replicates and quickly spreads to everyone in the household, so too do Legos, once even a single model enters the home, rapidly start spreading, covering every floor with multi-colored plastic." The Oxford research paves the way for work on preventive measures to halt Lego spread, allowing parents to once again walk barefoot in their homes without fear.

Mathematician **Dr. H. Leon Lilly** at M.I.T. for solving the Pee-ometry Equation. Considered perhaps the most important development in mathematics since Pythagoras and his triangle, the proof for this simple, elegant equation has immediate real-world applications, enabling parents in the midst of a diaper change to quickly calculate the trajectory that an infant boy will arc his urine once air hits his penis, and dodge the stream.

Neuroscientists **Drs. Kolora Flamebreech** and **Simbana Bam** at the University of New South Wales in Australia for identifying and validating the postpartum memory lapse—known colloquially as "pregnancy brain" before the baby arrives and "mom brain" after the birth. This will undoubtedly cheer millions of moms who'll be glad to know they're not losing their minds, just their short-term memories. Fortunately, these brain farts are like regular farts—annoying but transient. The upside, researchers report, is that this type of memory loss serves to erase any recollection of delivery-room trauma, with the result that moms actually (might) want to have sex again. (At some point. Probably.)

Acknowledgments

Generosity is the currency of the blogosphere, and we would not have gotten to the point of publishing this book were it not for the incredible camaraderie and support from all of the mom and dad bloggers in our little corner of the Internet. We are grateful for every Like, Share, Tweet, and Pin. Thank you!

Very special thanks to Tim Sullivan, Leon Lilly, Becky Blades, and the entire faculty of BlogU. Great big thank yous to Jill Smokler, Jen Mann, Nicole Knepper, Val Curtis, Leslie Marinelli, Deva Dalporto, Jerry Mahoney, Jason Good, Johanna Stein, D.J. Paris, Tracy Beckerman, and David Vienna for your counsel and advice and for letting us stand on your platforms so that Science of Parenthood could grow.

Thanks to Stephanie Jankowski, Colleen Gresh, and Cliff and Danielle Cagle for sharing their truly inspirational ideas. Thanks also to early readers Kevin Haynes, Arlana Vincent Guckenberger and Shari Dworkin-Smith who gave us fresh eyes and valuable insights.

Thanks to teacher Samella Mejia, who introduced Norine's son Fletcher to Isaac Newton's laws of force and motion, sparking the concept for Science of Parenthood in the process.

Many thanks to Brooke Warner, publisher of She Writes Press, who was there at the birth of Science of Parenthood and who believed in this book from the very beginning and held our hands all through the publishing process. We truly appreciate your support and encouragement. Thanks also to our publicist, Joanne McCall, who made sure that someone outside of our families knew we'd written this book.

Enormous thanks to our parents, Perry and Eileen Dworkin, angel investors and babysitters extraordinaire; and Jack and Jean Ziegler—to the former, for proving that following an insane dream isn't always a terrible idea, and to the latter, for demonstrating that there is nothing so inspiring as a really BIG project. And to all, for offering forty-plus years (gulp) of endless encouragement.

There are not thanks enough for our husbands, Stewart and Greg, who are the real science geeks in our families and who supplied us with countless ideas, fact-checked our work, and kept the kiddos out of traffic so we could nurture our "book baby." We love you bunches!

Finally, a thousand and more thanks to our most precious boys, Fletcher and Holden, who provide endless inspiration and laughter . . . sometimes intentionally.

— Norine & Jessica
November 2015

About the Authors

Norine Dworkin-McDaniel is co-creator and principal writer of Science of Parenthood, named one of Parenting.com's "blogs every parent should read." A longtime magazine writer, Norine has written for *Parents*, *American Baby*, *The Huffington Post* and just about every women's magazine you can buy at supermarket checkout. She has also contributed to several humor anthologies, including *Have Milk Will Travel: Adventures in Breastfeeding*. Her high school math teacher would be stunned to hear that she finally gets geometry. Sort of.

Jessica Ziegler is Science of Parenthood's co-creator, illustrator, web designer, and contributing writer. In her "off hours," she is the director of social web design for Vestor Logic and the writer and illustrator of a customizable children's book series at StoryTots.com. She was named a 2014 Humor Voice of the Year by BlogHer and SheKnows. Her writing and illustration have been published on *The Huffington Post*, InThePowderRoom.com, VEGAS.com, and in *Las Vegas Life* and *Las Vegas Weekly*. Surprisingly, her work remains unheralded by the scientific community.

Together, Jessica and Norine are the creators of the highly acclaimed *The Big Book of Parenting Tweets* and *The Bigger Book of Parenting Tweets*.

Find out more about Science of Parenthood at www.ScienceOfParenthood.com.

If you enjoyed this book, please consider
leaving a rating on Amazon.com
or Goodreads.com.

It helps indie publishers
tremendously, and we truly appreciate it!

THANK YOU!

Visit ScienceOfParenthood.com to learn more.